P9-BJU-553

# Introduction to Emergency Management

# Introduction to Emergency Management

George D. Haddow
Jane A. Bullock

An imprint of Elsevier Science
Amsterdam Boston Heidelberg London New York Oxford Paris
San Diego San Francisco Singapore Sydney Tokyo

Butterworth-Heinemann is an imprint of Elsevier Science.

Recognizing the importance of preserving what has been written, Elsevier Science prints its books on acid-free paper whenever possible.

**Library of Congress Cataloging-in-Publication Data**
Haddow, George D.
Introduction to emergency management/George D. Haddow, Jane A. Bullock.
p. cm.
ISBN 0-7506-7689-2 (alk. paper)
1. Emergency management. 2. Emergency management–United States. I. Bullock, Jane A. II. Title

HV551.2.H3 2003
363.34′8′0973–dc21

2003045314

**British Library Cataloguing-in-Publication Data**
A catalogue record for this book is available from the British Library.

The publisher offers special discounts on bulk orders of this book.
For information, please contact:
Manager of Special Sales
Elsevier Science
200 Wheeler Road
Burlington, MA 01803

Tel: 781-313-4700
Fax: 781-313-4882

For information on all Butterworth-Heinemann publications available, contact our World Wide Web home page at: http://www.bh.com.

10 9 8 7 6 5 4 3 2 1

Printed in the United States of America

# Contents

# Introduction

Disasters affect someone, somewhere, everyday. The management of the effects of disasters is called emergency management. This book is designed to provide the reader with a primer on the background, components, and operations of the emergency management discipline. The book describes current practices, strategies, and the key players in emergency management in the United States and around the world. The intent is to provide the reader with a working knowledge of how the functions of emergency management operate and the influence they can have on everyday life.

The capacity and capabilities for emergency management vary significantly throughout the world. The principal focus of this book is the emergency management system in the United States and the agency that leads this system, the Federal Emergency Management Agency (FEMA). We chose this focus because the U.S. system has evolved into one of the most recognized and copied systems in the world. The U.S. system has experienced every form of disaster: natural, manmade, and political. The lessons learned from these experiences, the changes made in response to these events, and how the system continues to evolve because of new threats provide a solid landscape to examine what is emergency management. The book also focuses on the U.S. system because of the experiences and knowledge of the authors of the book. In the end, most people would agree that FEMA is the most-recognized, best equipped, and best-funded emergency management system and, by all measures, is the leading emergency management organization in the world.

This book does not focus exclusively on FEMA, however. State and local emergency management organizations are the focus of many of the case studies, and their roles as partners with FEMA are discussed throughout the book. Under the U.S. Constitution, states are given responsibility for public health and safety. The federal government becomes involved at the request of states when they are overwhelmed or unable to fulfill this basic function. The federal government is also the source of most of the funding for public health and safety programs. This has resulted in a strong federal presence in emergency management. The competition for scarce resources and the immediate priorities of state and local governments have contributed to the creation of a strong federal role in emergency management. This trend may be changing, as is discussed in later chapters.

A comprehensive chapter is included that describes emergency management activities in the international sector. As the appreciation and value of the need for a robust emergency management system has grown, governments across the globe have focused more attention on this issue. The chapter includes a detailed case study of the response to the 2001 earthquake in Gujarat, India.

No discussion of emergency management in the United States can now be complete without talking about the changes caused by the terrorists' attacks on

September 11, 2001. Consideration of terrorism is included throughout the chapters of this book, and a single chapter discusses the emerging terrorist threat on the implications to the U.S. emergency management system.

A brief summary of the contents of the book's 10 chapters follows:

Chapter 1, The Historical Context of Emergency Management, includes a brief discussion of the historical, organizational, and legislative evolution of emergency management in the United States by tracing the major changes triggered by disasters or other human or political events.

Chapter 2, Natural and Technological Hazards and Risk Assessment, identifies and defines the hazards confronting emergency management.

Chapter 3, The Disciplines of Emergency Management: Mitigation, discusses what the function of mitigation is and what strategies and programs are applied by emergency management or other disciplines to reduce the effects of disaster events.

Chapter 4, The Disciplines of Emergency Management: Response, focuses on the essential functions and processes of responding to a disaster event.

Chapter 5, The Disciplines of Emergency Management: Recovery, describes the broad range of government and voluntary programs available to assist individuals and communities in rebuilding in the aftermath of a disaster.

Chapter 6, The Disciplines of Emergency Management: Preparedness, catalogues the broad range of programs and processes that constitute the preparedness function of modern emergency management.

Chapter 7, The Disciplines of Emergency Management: Communications, breaks from the more traditional approach to emergency management and focuses on why communications with the public, the media, and partners is critical to emergency management of the 21st century.

Chapter 8, International Disaster Management, provides an overview of current activity in emergency management through an examination of selected international organizations.

Chapter 9, Emergency Management and the New Terrorist Threat, describes how the events of September 11 have altered the traditional perceptions of emergency management.

Chapter 10, The Future of Emergency Management, provides insights, speculations, conclusions, and recommendations from the authors on where emergency management is headed or should be going in the future.

Our goal in writing this book was to provide readers with an understanding of emergency management, insight into how events have shaped the discipline, and thoughts about the future direction of emergency management. In the end, we hope it will educate, inform, and possibly encourage individuals to actively participate in the practice of emergency management in their professions and communities.

# Acknowledgments

This book could not have been completed without the assistance of two valuable groups of partners. First, the authors are grateful to the Institute for Crisis, Disaster and Risk Management at The George Washington University and its Co-Directors, Dr. Jack Harrald and Dr. Joseph Barbera, for their support. The Institute's Greg Shaw was helpful in many ways. Second, we want to acknowledge the contributions of a group of graduate students at the Institute for their hard work in conducting the preliminary research and drafting for this book. This group made significant contributions to the book and includes Kathryn A. Allen, James A. Cooke, Damon P. Coppola, Elizabeth N. Halford, Robert D. Hulshouser, Nadeem U. Khan, Mariette M. Larrick, Stephen Marquette, Michele R. Novack, George Nunez, John C. Peyrebrune, Jeannette C. Rood, Christian M. Salmon, Monica M. Severson, and Carmen A. Whitson. Two students deserve special acknowledgment for their interest, commitment, and quality of their efforts: Damon P. Coppola and Alfredo Lagos.

Finally, the authors wish to thank their respective spouses, Dick Bullock and Kim Haddow, for their enduring good humor and patience.

# 1. The Historical Context of Emergency Management

## INTRODUCTION

Emergency management has ancient roots. Early hieroglyphics depict cavemen trying to deal with disasters. The Bible speaks of the many disasters that befell civilizations. In fact, the account of Moses parting the Red Sea could be interpreted as the first attempt at flood control. As long as there have been disasters, individuals and communities have tried to do something about them; however, organized attempts at dealing with disasters did not occur until much later in modern history.

The purpose of this chapter is to discuss the historical, organizational, and legislative history of modern emergency management in the United States. Some of the significant events and people that have shaped the emergency management discipline over the years will be reviewed. Understanding the history and evolution of emergency management is important because at different times, the concepts of emergency management have been applied differently. The definition of emergency management can be extremely broad and all-encompassing. Unlike other more structured disciplines, it has expanded and contracted in response to events, Congressional desires, and leadership styles.

A simple definition is that *emergency management* is the discipline dealing with risk and risk avoidance. Risk represents a broad range of issues and includes an equally diverse set of players. The range of situations that could possibly involve emergency management or the emergency management system is extensive. This supports the premise that emergency management is integral to the security of everyone's daily lives and should be integrated into daily decisions and not just called on during times of disasters.

Emergency management is an essential role of government. The Constitution tasks the states with responsibility for public health and safety—hence responsible for public risks—with the federal government in a secondary role. The federal role is to help when the state, local, or individual entity is overwhelmed. This fundamental philosophy continues to guide the government function of emergency management.

Based on this strong foundation, the validity of emergency management as a government function has never been in question. Entities and organizations fulfilling the emergency management function existed at the state and local level long before the federal government became involved. But as events occurred, as political philosophies changed, and as the nation developed, the federal role in emergency management steadily increased.

## EARLY HISTORY: 1800–1950

In 1803, a Congressional Act was passed to provide financial assistance to a New Hampshire town that had been devastated by fire. This is the first example of the federal government becoming involved in a local disaster. It was not until the administration of Franklin Roosevelt began to use government as a tool to stimulate the economy that a significant investment in emergency management functions was made by the federal government.

During the 1930s, the Reconstruction Finance Corporation and the Bureau of Public Roads were both given authority to make disaster loans available for repair and reconstruction of certain public facilities after disasters. The Tennessee Valley Authority was created during this time to produce hydroelectric power and, as a secondary purpose, to reduce flooding in the region.

A significant piece of emergency management legislation was passed during this time. The Flood Control Act of 1934 gave the U.S. Army Corps of Engineers increased authority to design and build flood control projects. This act has had a significant and long-lasting impact on emergency management in this country. This act reflected a philosophy that man could control nature, thereby eliminating the risk of floods. Although this program would promote economic and population growth patterns along the nation's rivers, history has proven that this attempt at emergency management was shortsighted and costly.

## THE COLD WAR AND THE RISE OF CIVIL DEFENSE: 1950s

The next notable timeframe for the evolution of emergency management occurs during the 1950s. The era of the Cold War presented the principal disaster risk as the potential for nuclear war and nuclear fallout. Civil Defense programs proliferated across communities during this time. Individuals and communities were encouraged to build bomb shelters to protect themselves and their families from nuclear attack from the Soviet Union.

Almost every community had a civil defense director, and most states had someone who represented civil defense in their state government hierarchy. By profession, these individuals were usually retired military personnel, and their operations received little political or financial support from their state or local governments. Equally often, the civil defense responsibility was an addition to other duties.

Federal support for these activities was vested in the Federal Civil Defense Administration (FCDA), an organization with little staff or financial resources whose main role was to provide technical assistance. In reality, the local and state civil defense directors were the first recognized face of emergency management in the United States.

A companion office to the FCDA, the Office of Defense Mobilization was established in the Department of Defense (DOD). The primary functions of this office were to allow for quick mobilization of materials and production and stockpiling of critical materials in the event of a war. It included a function called *emergency*

**Figure 1-1** Midwest Floods, June 1994—Homes, businesses, and personal property were all destroyed by the high flood levels. A total of 534 counties in nine states were declared for federal disaster aid. As a result of the floods, 168,340 people registered for federal assistance. FEMA News Photo.

*preparedness*. In 1958, these two offices were merged into the Office of Civil and Defense Mobilization.

The 1950s were a quiet time for large-scale natural disasters. Hurricane Hazel, a Category 4 hurricane, inflicted significant damage in Virginia and North Carolina in 1954; Hurricane Diane hit several mid-Atlantic and northeastern states in 1955; and Hurricane Audrey, the most damaging of the three storms, struck Louisiana and North Texas in 1957. Congressional response to these disasters followed a familiar pattern of ad hoc legislation to provide increased disaster assistance funds to the affected areas.

As the 1960s started, three major natural disaster events occurred. In a sparsely populated area of Montana, the Hebgen Lake Earthquake, measuring 7.3 on the Richter scale, brought attention to the fact that the nation's earthquake risk went beyond the California borders. Also in 1960, Hurricane Donna hit the west coast of Florida, and Hurricane Carla blew into Texas in 1961. The incoming Kennedy administration decided to make a change to the federal approach. In 1961 it created the Office of Emergency Preparedness inside the White House to deal with natural disasters. Civil Defense responsibilities remained in the Office of Civil Defense within the DOD.

## NATURAL DISASTERS BRING CHANGES TO EMERGENCY MANAGEMENT: 1960s

As the 1960s progressed, the United States would be struck by a series of major natural disasters. The Ash Wednesday Storm in 1962 devastated more than 620 miles of shoreline on the East Coast, producing more than $300 million in damages. In 1964, an earthquake measuring 9.2 on the Richter scale in the Prince William Sound, Alaska, became front-page news throughout America and the world. This quake generated a tsunami that affected beaches as far down the Pacific Coast as California and killed 123 people. Hurricane Betsy struck in 1965, and Hurricane Camille in 1969, killing and injuring hundreds of people and causing hundreds of millions of dollars in damage along the Gulf Coast.

As with previous disasters, the response was passage of ad hoc legislation for funds; however, the financial losses resulting from Hurricane Betsy's path across Florida and Louisiana started a discussion of insurance as a protection against future floods and a potential method to reduce continued government assistance after disasters. Congressional interest was prompted by the unavailability of flood protection insurance on the standard homeowner policy. Where this type of insurance was available, it was cost prohibitive. These discussions eventually led to passage of the National Flood Insurance Act of 1968, which created the National Flood Insurance Program (NFIP).

Congressman Hale Boggs of Louisiana is appropriately credited with steering this unique legislation through Congress. Unlike previous emergency management/ disaster legislation, this bill sought to do something about the risk before the disaster struck. It brought the concept of *community-based mitigation* into the practice of emergency management. In simple terms, when a community joined the NFIP, in exchange for making federally subsidized, low-cost flood insurance available to its citizens, the community had to pass an ordinance restricting future development in its floodplains. The federal government also agreed to help local communities by producing maps of their community's floodplains.

The NFIP began as a voluntary program as part of a political compromise that Boggs negotiated. As a voluntary program, few communities joined. After Hurricane Camille struck the Louisiana, Alabama, and Mississippi coasts in 1969, the goals of the NFIP to protect people's financial investments and to reduce government disaster expenditures were not being met. It took Hurricane Agnes devastating Florida for a change to occur.

George Bernstein, brought down from New York Department of Insurance by President Nixon to run the Federal Insurance Administration (FIA) within the Department of Housing and Urban Development (HUD), proposed linking the mandatory purchase of flood insurance to all homeowner loans backed by federal mortgages. This change created an incentive for communities to join the NFIP because a significant portion of the home mortgage market was federally backed. This change became the Flood Insurance Act of 1972.

It is important to note how local and state governments chose to administer this flood risk program. Civil defense departments usually had responsibility to deal with risks and disasters. Although the NFIP dealt with risk and risk avoidance, responsibilities for the NFIP were sent to local planning departments and State Departments of Natural Resources. This reaction is one illustration of the fragmented and piecemeal approach to emergency management that evolved during the 1960s and 1970s.

## THE CALL FOR A NATIONAL FOCUS TO EMERGENCY MANAGEMENT: 1970s

In the 1970s, responsibility for emergency management functions was evident in more than five federal departments and agencies, including the Department of Commerce (weather, warning, and fire protection), the General Services Administration (continuity of government, stockpiling, federal preparedness), the Treasury Department (import investigation), the Nuclear Regulatory Commission (power plants), and HUD (flood insurance and disaster relief).

With passage of the Disaster Relief Act of 1974, prompted by the previously mentioned hurricanes and the San Fernando earthquake of 1971, HUD possessed the most significant authority for natural disaster response and recovery through the NFIP under the FIA and the Federal Disaster Assistance Administration (disaster response, temporary housing, and assistance). On the military side, there existed the Defense Civil Preparedness Agency (nuclear attack) and the U.S. Army Corps of Engineers (flood control); however, taking into account the broad range of risks and potential disasters, more than 100 federal agencies were involved in some aspect of risk and disasters.

This pattern continued down to the state and, to a lesser extent, local levels. Parallel organizations and programs added to confusion and turf wars, especially during disaster response efforts. The states and the governors grew increasingly frustrated over this fragmentation. In the absence of one clear federal lead agency in emergency management, a group of State Civil Defense Directors led by Lacy Suiter of Tennessee and Erie Jones of Illinois launched an effort through the National Governor's Association (NGA) to consolidate federal emergency management activities in one agency.

With the election of a fellow state governor, President Jimmy Carter of Georgia, the effort gained steam. President Carter came to Washington committed to streamlining all government agencies and seeking more control over key administrative processes. The state directors lobbied the NGA and Congress for a consolidation of federal emergency management functions. When the Carter administration proposed such an action, it met with a receptive audience in the Senate. Congress had already expressed concerns about the lack of a coherent federal policy and the inability of states to know whom to turn to in the event of an emergency.

The federal agencies involved were not as excited about the prospect. A fundamental law of bureaucracy is a continued desire to expand control and authority, not to lose control. In a consolidation of this sort, there would be losers and winners. There was a question of which federal department/agency should house the new consolidated structure. As the debate continued, the newly organized National Association of State Directors of Emergency Preparedness championed the creation of a new independent organization, an idea that was quickly supported by the Senate.

In the midst of these discussions, an accident occurred at the Three Mile Island Nuclear Power Plant in Pennsylvania, which added impetus to the consolidation effort. This accident brought national media attention to the lack of adequate off-site preparedness around commercial nuclear power plants and the role of the federal government in responding to such an event.

On June 19, 1978, President Carter transmitted to the Congress the Reorganization Plan Number 3 (3 CFR 1978, 5 U.S. Code 903). The intent of this plan was to consolidate emergency preparedness, mitigation, and response activities into one federal emergency management organization. The President stated that the plan would establish the Federal Emergency Management Agency (FEMA) and that the FEMA Director would report directly to the President.

Reorganization Plan Number 3 transferred the following agencies or functions to FEMA: National Fire Prevention Control Administration (Department of Commerce), Federal Insurance Administration (HUD), Federal Broadcast System (Executive Office of the President), Defense Civil Preparedness Agency (DOD), Federal Disaster Assistance Administration (HUD), and the Federal Preparedness Agency (GSA).

Additional transfers of emergency preparedness and mitigation functions to FEMA were: Oversight of the Earthquake Hazards Reduction Program (Office of Science and Technology Policy), coordination of dam safety (Office of Science and Technology Policy), assistance to communities in the development of readiness plans for severe weather-related emergencies, coordination of natural and nuclear disaster warning systems, and coordination of preparedness and planning to reduce the consequences of major terrorist incidents.

The plan articulated several fundamental organizational principles:

> First, Federal authorities to anticipate, prepare for, and respond to major civil emergencies should be supervised by one official responsible to the President and given attention by other officials at the highest levels. Second, an effective civil defense system requires the most efficient use of all available resources. Third, whenever possible, emergency responsibilities should be extensions of Federal agencies. Fourth, Federal hazard mitigation activities should be closely linked with emergency preparedness and response functions.

Subsequent to Congressional review and concurrence, the Federal Emergency Management Agency was officially established by Executive Order 12127 of March 31, 1979 (44 FR 19367, 3 CFR, Comp., p. 376). A second Executive Order, 12148, mandated reassignment of agencies, programs, and personnel into the new entity FEMA.

Creating the new organization made sense, but integrating the diverse programs, operations, policies, and people into a cohesive operation was a much bigger task than realized when the consolidation began. It would take extraordinary leadership and a common vision. The consolidation also created immediate political problems. By consolidating these programs and the legislation that created them, FEMA would have to answer to 23 Committees and Subcommittees in Congress with oversight of its programs. Unlike most other federal agencies, it would have no organic legislation to support its operations and no clear champions to look to during the Congressional appropriations process.

In addition, President Carter had problems finding a director for this new organization. No large constituent group was identified with emergency management. Furthermore, the administration was facing major problems with Congress and the public because of the Iranian hostage crisis. President Carter finally reached into his own cabinet and asked John Macy, then head of the Office of Personnel Management (OPM), to become Director of FEMA.

John Macy's task was to unify an organization that was not only physically separated—parts of the agency were located in five different buildings around Washington—but also philosophically separate. Programs focused on nuclear war preparations were combined with programs focused on a new consciousness of the environment and floodplain management. Macy focused his efforts by emphasizing the similarities between natural hazards preparedness and civil defense by developing a new concept called the Integrated Emergency Management System (IEMS). This system was an all-hazards approach that included direction, control, and warning as functions common to all emergencies from small, isolated events to the ultimate emergency of nuclear attack.

For all of his good efforts, FEMA continued to operate as individual entities pursuing their own interests and answering to their different Congressional bosses. It was a period of few major disasters, so virtually nobody noticed this problem of disjointedness.

## CIVIL DEFENSE REAPPEARS AS NUCLEAR ATTACK PLANNING: 1980s

The early and middle 1980s saw FEMA facing many challenges but no significant natural disasters. The absence of the need for a coherent federal response to disasters, as was called for by Congress when it approved the establishment of FEMA, allowed FEMA to continue to exist as an organization of many parts.

In 1982, President Reagan appointed Louis O. Guiffrida as Director of FEMA. Guiffrida, a California friend of Ed Meese, one of the President's closest advisors, had a background in training and terrorism preparedness at the state government level. Guiffrida proceeded to reorganize FEMA consistent with administration policies and his background. Top priority was placed on government preparedness for a nuclear attack. Resources within the agency were realigned, and additional budget authority was sought to enhance and elevate the national security responsibilities of the agency. With no real role for the states in these national security activities, the state directors who had lobbied for the creation of FEMA saw their authority and federal funding declining.

Guiffrida also angered one of the only other visible constituents of the agency—the fire services community. Guiffrida diminished the authority of the U.S. Fire Administration by making it part of FEMA's Directorate of Training and Education. The newly acquired campus at Emmitsburg, Maryland, was intended to become the preeminent National Emergency Training Center (NETC).

During Guiffrida's tenure, FEMA faced several unusual challenges that stretched its authority, including asserting FEMA into the lead role for continuity of civilian government in the aftermath of a nuclear attack, managing the federal response to the contamination at Love Canal and Times Beach, Missouri, and the Cuban refugee crisis. Although Guifridda managed to bring the agency physically together in a new headquarters building in Southwest Washington, severe morale problems persisted.

Dislike of Guiffrida's style and questions about FEMA's operations came to the attention of U.S. Representative Al Gore of Tennessee, who then served on the

House Science and Technology Committee. As the Congressional hearings proceeded, the Department of Justice and a grand jury began investigations of senior political officials at FEMA. These inquiries led to the resignation of Guiffrida and top aides in response to a variety of charges, including misuse of government funds, but the shakeup marked a milestone of sorts: FEMA and emergency management had made it into the comic strip "Doonesbury."

President Reagan then selected General Julius Becton to be Director of FEMA. General Becton was a retired military general and had been the Director of the Office of Foreign Disaster Assistance in the State Department. General Becton is uniformly credited with restoring integrity to the operations and appropriations of the agency. From a policy standpoint, he continued to emphasize the programs of his predecessor but in a less visible manner. Becton expanded the duties of FEMA when he was asked by the DOD to take over the program dealing with the off-site cleanup of chemical stockpiles on DOD bases. This program was fraught with problems, and bad feelings existed between the communities and the bases over the funds available to the communities for the cleanup. FEMA had minimal technical expertise to administer this program and was dependent on the DOD and the Army for the funding. This situation led to political problems for the agency and did not lead to significant advancements in local emergency management operations, as promised by the DOD.

At one point in his tenure, General Becton ranked the programs in FEMA by level of importance. Of the more than 20 major programs, the earthquake, hurricane, and flood programs ranked near the bottom. This priority seems logical based on the absence of any significant natural hazards, but this situation is noteworthy in the context that it continued the pattern of isolating resources for national security priorities without recognizing the potential of a major natural disaster.

This issue was raised by then Senator Al Gore in hearings on FEMA's responsibilities as lead agency for the National Earthquake Hazards Reduction Program (NEHRP). Senator Gore, reacting to a scientific report that said there could be up to 200,000 casualties from an earthquake occurring on the New Madrid fault, believed that FEMA's priorities were misplaced. The legislation that created the NEHRP called on FEMA to develop a plan for how the federal government would respond to a catastrophic earthquake. This Federal Response Plan would later become the operating Bible for all of the federal agencies response operations. Senator Gore concluded that FEMA needed to spend more time working with its federal, state, and local partners on natural hazards planning.

## AN AGENCY IN TROUBLE: 1989–1992

As Congress debated, and finally passed, major reform of federal disaster policy as part of the Stewart McKinney–Robert T. Stafford Act, the promise of FEMA and its ability to support a national emergency management system remained in doubt.

As the 1980s closed, FEMA was an agency in trouble. It suffered from severe morale problems, disparate leadership, and conflicts with its partners at the state and local level over agency spending and priorities. In 1989, two devastating natural disasters called the continued existence of FEMA into question. In September, Hurricane

Hugo slammed into North Carolina and South Carolina after first hitting Puerto Rico and the Virgin Islands. It was the worst hurricane in a decade, with more than $15 billion in damages and 85 deaths. FEMA was slow to respond, waiting for the process to work and for the governors to decide what to do. Sen. Ernest Hollings (D-SC) personally called the FEMA Director and asked for help, but the Agency moved slowly. Hollings went on national television to berate FEMA in some of the most colorful language ever, calling the agency the "sorriest bunch of bureaucratic jackasses."

Less than a month later, the Bay Area of California was rocked by the Loma Prieta Earthquake as the 1989 World Series got under way in Oakland Stadium. FEMA was not prepared to respond, but it was lucky. Although FEMA had spent the last decade focused on nuclear attack planning, FEMA's state partners in emergency management, especially in California, had been preparing for a more realistic risk—an earthquake. Damages were high, but few lives were lost. This outcome was a testament to good mitigation practices in building codes and construction that were adopted in California and some good luck relative to the time when the earthquake hit.

A few years later, FEMA was not so lucky. In August 1992, Hurricane Andrew struck Florida and Louisiana, and Hurricane Iniki struck Hawaii within months of each other. FEMA wasn't ready, and neither were FEMA's partners at the state level. The agency's failure to respond was witnessed by Americans all across the country as major news organizations followed the crisis. The efficacy of FEMA as the national emergency response agency was in doubt. President Bush dispatched then Secretary of Transportation Andrew Card to take over the response operation and sent in the military.

It was not just FEMA that failed in Andrew; it was the process and the system. In Hurricane Andrew, FEMA recognized the need to apply all of its resources to the response and began to use its national security assets for the first time in a natural disaster response—but it was too late. Starting with Hurricane Hugo, public concern over natural disasters was high. People wanted, and expected, government to be there to help in their time of need. FEMA seemed incapable of carrying out the essential government function of emergency management.

In the aftermath of Hurricanes Andrew and Iniki, there were calls for abolishing FEMA. Investigations by the General Accounting Office (GAO) and other government and nongovernmental watchdog groups called for major reforms. None of this was lost on the incoming Clinton administration. As Governor of Arkansas, President Clinton had experience responding to several major flooding disasters and realized how important an effective response and quick recovery was to communities and to voters. At his side throughout these disasters was James Lee Witt, former County Judge and Administrator of Yell County, and, later, the State Director for Emergency Management in Arkansas.

## THE WITT REVOLUTION: 1993–2001

When President Clinton nominated James Lee Witt to be Director of FEMA, he breathed life back into FEMA and brought a new style of leadership to the troubled agency. Witt was the first Director of FEMA with emergency management

**Figure 1-2** Northridge Earthquake, CA, January 17, 1994—Many roads, including bridges and elevated highways were damaged by the 6.7 magnitude earthquake. Approximately 114,000 residential and commercial structures were damaged and 72 deaths were attributed to the earthquake. Damage costs were estimated at $25 billion. FEMA News Photo.

experience. He was from the constituency who had played a major role in creating FEMA but had been forgotten, the state directors. With Witt, President Clinton had credibility and, more important, a skilled politician who knew the importance of building partnerships and serving customers.

Witt came in with a mandate to restore the trust of the American people that their government would be there for them during times of crisis. He initiated sweeping reforms inside and outside the agency. Inside FEMA, he reached out to all employees, implemented customer service training, and reorganized the agency to break down bottlenecks. He supported application of new technologies to the delivery of disaster services and focused on mitigation and risk avoidance. Outside of the agency, he strengthened the relationships with state and local emergency managers and built new ones with Congress, within the administration, and with the media. Open communication internally and externally was one of the hallmarks of the Witt years at FEMA.

Witt's leadership and the changes he made were quickly tested as the nation experienced an unprecedented series of natural disasters. The Midwest floods in 1993 resulted in major disaster declarations in nine states. The Midwest floods called into question the value of some of the flood control measures initiated long ago as part of the 1930s Army Corps of Engineers' legislation. FEMA's successful response to these floods brought the opportunity to change the focus of post-disaster recovery by initiating the largest voluntary buyout and relocation program to date in an effort to move people out of the floodplain and out of harm's way.

The Northridge, California, earthquake quickly followed the Midwest floods in 1994. Northridge tested all of the new streamlined approaches and technology

**Figure 1-3** Franklin, VA, September 21, 1999—Hurricane Floyd left the downtown section of Franklin, VA, under six feet of water. The water has begun to recede, as shown by the high-water marks, but hazards still include propane tanks, gas tanks, chemical barrels, and pesticides. Photo by Liz Roll/FEMA News Photo.

advancements for delivery of services and created some more. Throughout the next several years, FEMA and its state and local partners would face every possible natural hazard, including killer tornadoes, ice storms, hurricanes, floods, wildfires, and drought.

When President Clinton elevated Witt as Director of FEMA to be a member of his Cabinet, the value and importance of emergency management were recognized. Witt used this promotion as an opportunity to lobby the nation's governors to include their state emergency management directors in their Cabinets.

The Oklahoma City Bombing in April 1995 represented a new phase in the evolution of emergency management. This event, following after the first bombing of the World Trade Center in New York City in 1992, raised the issue of America's preparedness for terrorism events. Because emergency management responsibilities are defined by risks and the consequences of those risks, responding to terrorist threats was included. The Oklahoma City bombing tested this thesis and set the stage for interagency disagreements over which agency would be in charge of terrorism.

The Nunn-Lugar legislation of 1995 left the question open as to who would be the lead agency in terrorism. Many people fault FEMA leadership for not quickly claiming that role, and the late 1990s were marked by several different agencies and departments having a role in terrorism planning. The question of who is the first responder to a terrorism incident—fire, police, emergency management, or emergency medical services—was closely examined without any clear answers. The state directors were looking for FEMA to claim the leadership role. In an uncharacteristic way, the leadership of FEMA vacillated on this issue. Terrorism was certainly part of the all-hazards approach to emergency management championed by FEMA, but the resources and technologies needed to address specific issues such as biochemical warfare and weapons of mass destruction events seemed well beyond the reach of the current emergency management structure.

While this debate continued, FEMA took an important step in its commitment to disaster mitigation by launching a national initiative to promote a new community-based approach called Project Impact: Building Disaster-Resistant Communities. This project was designed to mainstream emergency management and mitigation practices into every community in America. It went back to the roots of emergency management. It asked a community to identify risks and establish a plan to reduce those risks. It asked communities to establish partnerships that included all of the stakeholders in the community, including, for the first time, the business sector.

The goal of Project Impact was to incorporate decisions about risk and risk avoidance into the community's everyday decision-making processes. By building a disaster-resistant community, the community would promote sustainable economic development, protect and enhance its natural resources, and ensure a better quality of life for its citizens. Project Impact had ambitious goals and was well received by the communities and Congress. It was designed to create a broader constituency, a grassroots campaign, for emergency management issues.

As the decade ended without any major technological glitches from Y2K, FEMA was recognized as the preeminent emergency management system in the world. It was emulated in other countries, and Witt became an ambassador for emergency management overseas. Hurricane Mitch saw a change in American foreign policy toward promoting and supporting community-based mitigation projects. State and local emergency management programs had grown and their value recognized and supported by society. Private-sector and business continuity programs were flourishing.

The role and responsibility and the partnerships supporting emergency management had significantly increased, and its budget and stature had grown. Good emergency management became a way to get economic and environmental issues on the table; it became a staple of discussion relative to a community's quality of life.

The profession of emergency management was attracting a different type of individual. Political and management skills were critical, and candidates for state, local, and private emergency management positions were now being judged on their training and experience rather than their relationship to the community's political leadership. Undergraduate and advance degree programs in emergency management were flourishing at more than 65 national colleges and universities. It was now a respected, challenging, and sought-after profession.

## TERRORISM BECOMES MAJOR FOCUS: 2001

With the election of George W. Bush, a new FEMA Director, Joe Allbaugh, was named to head the agency. As a former Chief of Staff to Governor Bush in Texas and President Bush's Campaign Manager in the 2000 presidential race, Allbaugh had a close personal relationship with the President. As demonstrated by Witt and Clinton, this was viewed as a positive for the agency. His lack of emergency management background was not discussed as an issue during his confirmation hearings.

Allbaugh got off to a rocky start when the administration decided to eliminate funding for the popular Project Impact. Immediately after this decision was announced, the 6.8 magnitude Nisqually Earthquake shook Seattle, Washington. Seattle happened to be one of the most successful Project Impact communities. The Mayor of Seattle went on national television and credited Project Impact as responsible for why there was almost no damage from the quake. Later that evening, Vice President Dick Cheney was asked why the program was being eliminated and replied by saying that its effectiveness had been questioned. As FEMA's budget proceeded through the appropriations process, Congress put funding back into Project Impact.

As part of major reorganization of the agency, Allbaugh recreated the Office of National Preparedness (ONP). This office was first established in the 1980s during the Guiffrida reign for planning for World War III, but was eliminated by Witt in 1992. This action raised some concerns among FEMA's constituents and FEMA staff; however, this time the mission of the office focuses on terrorism.

In a September 10, 2001, speech, Director Allbaugh talked about his priorities as being firefighters, disaster mitigation, and catastrophic preparedness. These words seem prophetic in light of the events of September 11. As the events of that day unfolded, FEMA activated the Federal Response Plan, and response operations proceeded as expected in New York and Virginia. Most of the agency's senior leaders, including the Director, were in Montana, attending the Annual Meeting of the National Emergency Management Association (NEMA) that represents State Emergency Management Directors. The strength of the system was proven as hundreds of response personnel initiated operations within minutes of the events.

## THE FUTURE: 2003 AND BEYOND

In the aftermath of the terrorist attacks on September 11, FEMA and the newly formed Department of Homeland Security, together with their partners in emergency management, fire, police, and public health at the state and local government levels, have been charged with expanding and enhancing America's emergency management system. In the coming years, billions of dollars will be allocated from the federal government to state and local governments in order to expand existing programs and establish new ones designed to meet the new terrorism threat.

As the environment of emergency management has grown, the quality, skill base, technical demands, and caliber of its practitioners have increased. Terrorism provides another opportunity to expand this base. The goal of this book is to provide the reader with the background and working knowledge of disciplines of emergency management and how they can be applied to any profession and in everyday life.

In conclusion, as with previous defining events, the environment for emergency management will absorb the event and evolve to reflect its impacts. If history repeats itself, it will shift the focus to a more national approach to the problem and emphasize preparedness through training and equipment. The resiliency of the system allows for these midstream corrections. The long-term viability and measure of the influence of emergency management will continue to depend on its value to all citizens in all communities every day, not just during times of crisis.

# 2. Natural and Technological Hazards and Risk Assessment

## INTRODUCTION

A *hazard* is defined as a "source of danger that may or may not lead to an emergency or disaster and is named after the emergency/disaster that could be so precipitated." *Risk* is defined as "susceptibility to death, injury, damage, destruction, disruption, stoppage and so forth." *Disaster* is defined as an "event that demands substantial crisis response requiring the use of governmental powers and resources beyond the scope of one line agency or service" (National Governors Association, 1982).

Hazard identification is the foundation of all emergency management activities. When hazards react with the human or built environments, the risks associated with that hazard can be assessed. Understanding the risk posed by identified hazards is the basis for preparedness planning and mitigation actions. Risk, when realized, such as in the event of an earthquake, tornado, flood, and so on, becomes a disaster that prompts emergency response and recovery activities. All emergency management activities are predicated on the identification and assessment of hazards and risks.

This chapter discusses the full range of existing hazards, both natural and technological. For each hazard, a brief description of the hazard and its effects is provided. Also included in this chapter is a discussion of risk assessment.

Much of the information for this chapter was acquired from the U.S. Federal Emergency Management Agency Website, www.fema.gov, and also from FEMA's book *Multi-Hazard Identification and Risk Assessment: A Cornerstone of the National Mitigation Strategy*. Included in Appendix B are organizations' Website addresses to reference for more in-depth information on a particular hazard.

## NATURAL HAZARDS

Natural hazards are those hazards that exist in the natural environment and pose a threat to human populations and communities. Human development has often exacerbated natural hazards. Building communities in the floodplain or on barrier islands increases the potential damage caused by flooding and storm surge. Building a school on a known earthquake fault increases the potential that the school will be destroyed by an earthquake. How humans can better live with hazards is the principal topic of Chapter 3, "The Disciplines of Emergency Management: Mitigation."

## Floods

Floods can be slow- or fast-rising but generally develop over a period of days. Floods usually occur from large-scale weather systems generating prolonged rainfall or onshore winds. Other causes of flooding include locally intense thunderstorms, snowmelt, ice jams, and dam failures. Floods are capable of undermining buildings and bridges, eroding shorelines and riverbanks, tearing out trees, washing out access routes, and causing loss of life and injuries. Flash floods usually result from intense storms dropping large amounts of rain within a brief period. Flash floods occur with little or no warning and can reach full peak in only a few minutes (Figure 2-1).

Floods are the most frequent and widespread disaster in many countries around the world. Historically, human development has congregated around rivers and ports, and transportation of goods has most commonly been conducted by water. This relationship has resulted in greater exposure to floods. For example, FEMA estimates that more than 9 million households and $390 billion in property are at risk from flooding in the United States alone. Flood losses paid by FEMA's National Flood Insurance Program in the 1990s totaled in the billions of dollars (Table 2-1).

Governments in many countries maintain river and stream gauges to measure floodwater elevations and to provide information on rising water for use in sandbagging and dyke construction and to warn populations of an impending flood.

**Figure 2-1** Midwest Floods, June 1994. Homes, businesses, and personal property were all destroyed by the high flood levels. A total of 534 counties in nine states were declared, eligible for federal disaster aid. As a result of the floods, 168,340 people registered for federal assistance. FEMA News Photo.

**Table 2-1** Top Ten U.S. Flood Disasters, 1990–2001

| Event | Date | # of Paid Losses | Amount of Paid Losses |
|---|---|---|---|
| Tropical Storm Allison | June 2001 | 30,031 | $1,062,182,904 |
| Louisiana Flood | May 1995 | 31,262 | $584,029,961 |
| Hurricane Floyd | September 1999 | 18,566 | $436,574,959 |
| Nor'easter | December 1992 | 24,676 | $341,862,939 |
| Hurricane Opal | October 1995 | 9,904 | $398,784,840 |
| Midwest Floods | June 1993 | 10,257 | $271,326,881 |
| Texas Flood | October 1994 | 6,153 | $216,632,740 |
| Hurricane Fran | September 1996 | 9,879 | $213,571,284 |
| March Storm | March 1993 | 9,631 | $210,833,769 |
| Northeast Flood | January 1996 | 11,679 | $175,026,721 |

*Source:* FEMA, www.fema.gov

## THE GREAT MIDWEST FLOODS OF 1993: RECOVERY COSTS

- A total of 534 counties in nine states were declared eligible for federal disaster aid for the 1993 Midwest Floods. As a result of the floods, 168,340 people registered for federal assistance.
- According to the Galloway Report in June 1994, estimated federal response and recovery costs included more than $4.2 billion in direct federal assistance, $1.3 billion in federal flood insurance payments, and more than $621 million in federal loans to individuals, businesses, and communities.

Of those totals, an estimated $1.69 billion was provided by the USDA for food stamps/commodities, crop loss payments, and other emergency farm grant and loan programs; $597 million by SBA for loans to homeowners, renters, and businesses; $500 million by HUD for housing and community grants; $200 million by DOC for economic development programs; $253 million by USACE for flood control and other emergency operations; $75 million by HHS for various public health services; $100 million by DOEd for schools and student aid; $64.6 million by DOL for employment training and temporary job assistance; $146.7 million by DOT for federal highway repairs, rail freight assistance, and other transportation and emergency services; $34 million by EPA for environmental abatement, control, and cleanup projects; and $41.2 million by DOI for various construction, survey, and cultural restoration programs.

FEMA's costs currently total $1.17 billion, including $371 million in grants to individuals and families for temporary housing, home repairs, unemployment payments, and other disaster-related expenses; $539.5 million to states and local governments for public property restoration and cleanup work; $167.6 million for property acquisitions and other hazard mitigation projects; and $29.2 million to other federal agencies for delivery of emergency supplies and other mission-assigned work.

*Note:* All funding amounts are in current CY2000 dollars, unadjusted for inflation.
*Source:* FEMA, www.fema.gov

## Earthquakes

An earthquake is a sudden, rapid shaking of the earth caused by the breaking and shifting of rock beneath the earth's surface. This shaking can cause buildings and bridges to collapse; disrupt gas, electric, and phone service; and sometimes trigger landslides, avalanches, flash floods, fires, and huge, destructive ocean waves (tsunamis). Buildings with foundations resting on unconsolidated landfill, old waterways, or other unstable soil are most at risk. Earthquakes can occur at any time of the year (Figure 2-2).

Specific active seismic zones have been identified around the globe. Millions of people live in these seismic zones and are exposed to the threat of an earthquake daily. The damage caused by an earthquake can be extensive, especially to incompatible building types and construction techniques. Also, earthquakes usually ignite fires, which can spread rapidly among damaged buildings if the water system has been disabled and fire services cannot access the site of the fire. Thousands of residents of Kobe, Japan, perished in the fires caused by the 1995 earthquake in that city because fire trucks and personnel were unable to get to the fires because of debris from fallen and damaged buildings blocking the streets.

Earthquakes are sudden events despite scientists' and soothsayers' best efforts to predict when they will occur. Seismic sensing technology can track seismic activity but has yet to accurately predict when a major seismic shift will occur that causes an earthquake. The effects of earthquakes are commonly described by the Richter scale.

**Figure 2-2** Northridge Earthquake, California, January 17, 1994. Buildings, cars, and personal property were all destroyed when the earthquake struck. Approximately 114,000 residential and commercial structures were damaged and 72 deaths were attributed to the earthquake. Damage costs were estimated at $25 billion. FEMA News Photo.

**Table 2-2** Estimated Earthquake Losses, 1987–1997

| Date | Location | Amount |
|---|---|---|
| November 24, 1987 | Southern California | $4 million |
| October 18, 1989 | Northern California | $5.6 million |
| February 28, 1990 | Southern California | $12.7 million |
| April 25, 1991 | Northern California | $66 million |
| June 28, 1992 | Southern California | $92 million |
| January 17, 1994 | Southern California | $13–20 billion |

*Source:* United States Geological Survey (USGS)

## THE RICHTER SCALE

The Modified Mercalli Intensity scale also measures the effects of earthquakes. The intensity of a quake is evaluated according to the observed severity of the quake at specific locations. The Mercalli scale rates the intensity on a Roman numeral scale that ranges from I to XII.

| Modified Mercalli | Damage Sustained | Richter Scale |
|---|---|---|
| I–IV Incidental to Moderate | No damage | < 4.3 |
| V Rather Strong | Damage negligible. Small, unstable objects displaced or upset; some dishes and glass broken. | 4.4–4.8 |
| VI Strong | Damage slight. Windows, dishes, glassware broken. Furniture moved or overturned. Weak plaster and masonry cracked. | 4.9–5.4 |
| VII Very Strong | Damage slight to moderate in well-built structures; considerable in poorly built structures. Furniture and weak chimneys broken. Masonry damaged. Loose bricks, tiles, plaster, and stones will fall. | 5.5–6.1 |
| VIII Destructive | Structural damage considerable, particularly to poorly built structures. Chimneys, monuments, towers, elevated tanks may fail. Frame houses moved. Trees damaged. Cracks in wet ground and steep slopes. | 6.2–6.5 |

*continues*

| | | |
|---|---|---|
| IX Ruinous | Structural damage severe; some buildings will collapse. General damage to foundations. Serious damage to reservoirs. Underground pipes broken. Conspicuous cracks in ground; liquefaction. | 6.6–6.9 |
| X Disastrous | Most masonry and frame structures/ foundations destroyed. Some well-built wooden structures and bridges destroyed. Serious damage to dams, dikes, embankments. Sand and mud shifting on beaches and flat land. | 7.0–7.3 |
| XI Very Disastrous | Few or no masonry structures remain standing. Bridges destroyed. Broad fissures in ground. Underground pipelines completely out of service. Widespread earth slumps and landslides. | 7.4–8.1 |
| XII Catastrophic | Damage nearly total. Large rock masses displaced. Lines of sight and level distorted. | > 8.1 |

*Source:* FEMA, www.fema.gov

## Hurricanes

All hurricanes start as tropical waves that grow in intensity and size to tropical depressions, which in turn grow to be tropical storms. A tropical storm is a warm-core tropical cyclone in which the maximum sustained surface wind speed ranges from 39 miles per hour (mph) to less than 74 mph. Tropical cyclones are defined as a low-pressure area of closed-circulation winds that originates over tropical waters. Winds rotate counterclockwise in the Northern Hemisphere and clockwise in the Southern Hemisphere.

A hurricane is a tropical storm with winds that have reached a constant speed of 74 mph or more. Hurricane winds blow in a large spiral around a relatively calm center known as the "eye." The eye is generally 20 to 30 miles wide, and the storm may extend outward for 400 miles. As a hurricane approaches, the skies will begin to darken and winds will strengthen. As a hurricane nears land, it can bring torrential rains, high winds, and storm surges. A single hurricane can last for more than two weeks over open waters and can run a path across the entire length of the eastern seaboard (Figure 2-3).

Hurricane season runs annually from June 1 through November 30. August and September are peak months during the hurricane season. Hurricanes are commonly described using the Saffir-Simpson scale.

**Figure 2-3** Hurricane Andrew, Florida, August 24, 1992. An aerial view showing damage from one of the most destructive hurricanes in America's history. One million people were evacuated and 54 died in this hurricane. FEMA News Photo.

## THE SAFFIR-SIMPSON SCALE

1 Wind Speed: 74–95 mph
Storm Surge: 4–5 feet above normal
Primary damage to unanchored mobile homes, shrubbery, and trees. Some coastal flooding and minor pier damage. Little damage to building structures.

2 Wind Speed: 96–110 mph
Storm Surge: 6–8 feet above normal
Considerable damage to mobile homes, piers, and vegetation. Coastal and low-lying area escape routes flood 2–4 hours before arrival of hurricane center. Buildings sustain roofing material, door, and window damage. Small craft in unprotected mooring break moorings.

3 Wind Speed: 111–130 mph
Storm Surge: 9–12 feet above normal
Mobile homes destroyed. Some structural damage to small homes and utility buildings. Flooding near coast destroys smaller structures; larger structures damaged by floating debris. Terrain continuously lower than 5 feet above sea level (ASL) may be flooded up to 6 miles inland.

4 Wind Speed: 131–155 mph
Storm Surge: 13–18 feet above normal
Extensive curtainwall failures, with some complete roof structure failure on small residences. Major erosion of beaches. Major damage to lower floors of structures near the shore. Terrain continuously lower than 10 feet ASL may flood (and require mass evacuations) up to 6 miles inland.

*continues*

**5** Wind Speed: Over 155 mph
Storm Surge: Over 18 feet above normal
Complete roof failure on many homes and industrial buildings. Some complete building failures. Major damage to lower floors of all structures located less than 15 feet ASL and within 500 yards of the shoreline. Massive evacuation of low ground residential areas may be required.

*Source:* FEMA

**Table 2-3** Top Ten Costliest Hurricanes in the United States, 1900–1996

| Hurricane | Year | Category | Damage |
|---|---|---|---|
| 1. Andrew (SE FL) | 1992 | 4 | $26 billion |
| 2. Hugo (SC) | 1989 | 4 | $7 billion |
| 3. Fran (NC) | 1996 | 3 | $3.2 billion |
| 4. Opal (NW FL/AL) | 1995 | 3 | $3 billion |
| 5. Frederic (AL/MS) | 1979 | 3 | $2.3 billion |
| 6. Agnes (NE U.S.) | 1972 | 1 | $2.1 billion |
| 7. Alicia (N TX) | 1983 | 3 | $2 billion |
| 8. Bob (NC and NE U.S.) | 1991 | 2 | $1.5 billion |
| 9. Juan (LA) | 1985 | 1 | $1.5 billion |
| 10. Camille (MS/AL) | 1969 | 5 | $1.42 billion |

*Source:* NOAA National Hurricane Center, www.nhc.noaa.gov

Hurricanes are capable of causing great damage and destruction over vast areas. Hurricane Floyd in 1999 first threatened the states of Florida and Georgia, made landfall in North Carolina, and damaged sections of South Carolina, North Carolina, Virginia, Maryland, Delaware, New Jersey, New York, Connecticut, Massachusetts, and Maine. The damage was so extensive in each of these states that they all qualified for federal disaster assistance. More recently, Hurricane Mitch brought death and destruction to Nicaragua, Guatemala, El Salvador, and Honduras.

In recent years, significant advances have been made in hurricane tracking technology and computer models. The National Hurricane Center in Miami, Florida, now tracks tropical waves from the moment they form off the coast of West Africa through their development as a tropical depression. Once the tropical depression grows to the strength of a tropical storm, the Hurricane Center assigns the storm a name. Once sustained wind speed of the tropical storm exceeds 74 mph, it becomes a hurricane. The Hurricane Center uses aircraft to observe and collect meteorological data on the hurricane and to track its movements across the Atlantic Ocean. It also uses several sophisticated computer models to predict the storm's path. These predictions are used by local and state emergency officials to make evacuation decisions and to predeploy response and recovery resources.

Historically, storm surge and high winds have been the principal contributors to the loss of life and injuries and the property and infrastructure damage caused by hurricanes. In recent years, inland flooding caused by hurricane rainfall has resulted in loss of life and severe property damage. Hurricanes also cause significant damage to the natural environment. Storm surge from hurricanes can result in severe beach erosion on barrier islands. Inland flooding from Hurricane Floyd inundated waste ponds on hog farms in North Carolina, washing the hog waste into the Cape Fear River, which eventually dumped these materials into the ocean.

## Storm Surges

Storm surges are storms that generate large waves on the coast that cause coastal flooding and erosion. They are most common from late fall to early spring but can develop year-round. They are usually associated with extra-tropical cyclones (Nor'easters) in the North Atlantic Ocean and the Gulf of Mexico, and severe winter low-pressure systems in the North Pacific Ocean and the Gulf of Alaska.

## Tornadoes

A tornado is a rapidly rotating vortex or funnel of air extending groundward from a cumulonimbus cloud. Approximately 1,000 tornadoes are spawned by thunderstorms each year. Most tornadoes remain aloft, but the danger is when they touch the ground. A tornado can lift and move huge objects, destroy or move whole buildings long distances, and siphon large volumes from bodies of water. Tornadoes follow the path of least resistance. People living in valleys have the greatest exposure to damage. Tornadoes are commonly described using the Fujita-Pearson Tornado scale.

### THE FUJITA-PEARSON TORNADO SCALE

F-0: 40–72 mph, chimney damage, tree branches broken
F-1: 73–112 mph, mobile homes pushed off foundation or overturned
F-2: 113–157 mph, considerable damage, mobile homes demolished, trees uprooted
F-3: 158–205 mph, roofs and walls torn down, trains overturned, cars thrown
F-4: 206–260 mph, well-constructed walls leveled
F-5: 261–318 mph, homes lifted off foundation and carried considerable distances, autos thrown as far as 100 meters

In the United States, the most susceptible states to tornadoes are Texas, Oklahoma, Arkansas, Missouri, and Kansas. Together these states occupy what is commonly known as "tornado alley." In recent years, however, tornadoes have struck in cities that are not regularly frequented by tornadoes, including Miami, Nashville, and Washington, D.C. Tornado season is generally March through August, although tornadoes can occur at any time of year. They tend to occur in the afternoons and evenings: more than 80 percent of all tornadoes strike between noon and midnight.

Tornadoes can have winds of up to 300 mph and possess tremendous destructive force. Damage is only incurred when the tornado touches down, but tornadoes can touch down in more than one place. The tornado that struck the Washington, D.C., metropolitan area in 2001 first touched down in Alexandria, Virginia, just south of the District of Columbia, went airborne over the District, and touched down again in College Park, Maryland, just north of the District (Figure 2-4).

Building collapse and flying debris are the principal causes of death and injuries by tornadoes. Early warning is the key to surviving in the path of a tornado. Doppler radar and other meteorological tools are improving the amount of advance warning time available before a tornado strikes. Improved communications and new technologies have also been critical to giving people advance warning of a tornado.

Buildings that are directly in the path of a tornado have little chance of surviving; however, new "safe room" technology developed by FEMA and Texas A&M University offers families and communities a method for surviving the tornado even if your home or community facility does not. A safe room can be built into an existing or new home for a small cost (estimated between $3,000 to $5,000) that will survive a tornado's high winds and flying debris. Your home may be destroyed, but anyone in the safe room will survive. Similar technology is being developed for community shelters.

Although reducing the loss of life and injuries is the principal goal of tornado preparedness and mitigation activities, new technologies in building design and construction are being developed by FEMA and others to reduce the damage to buildings and structures not located directly in the path of a tornado. Some of the

**Figure 2-4** College Park, Maryland, September 25, 2001. Rescue workers clean up the debris left by the tornado that killed two people and left more than $16.5 million in damages. Photo by Jocelyn Augustino/FEMA News Photo.

**Table 2-4** The 25 Deadliest U.S. Tornadoes

| Date | Place | Deaths |
|---|---|---|
| 1. March 18, 1925 | Tri-State (MO, IL, IN) | 689 |
| 2. May 6, 1840 | Natchez, MS | 317 |
| 3. May 27, 1896 | St. Louis, MO | 255 |
| 4. April 5, 1936 | Tupelo, MS | 216 |
| 5. April 6, 1936 | Gainesville, GA | 203 |
| 6. April 9, 1947 | Woodward, OK | 181 |
| 7. April 24, 1980 | Amite, LA; Purvis, MS | 143 |
| 8. June 12, 1899 | New Richmond, WI | 117 |
| 9. June 8, 1953 | Flint, MI | 115 |
| 10. May 11, 1953 | Waco, TX | 114 |
| 11. May 18, 1902 | Goliad, TX | 114 |
| 12. March 23, 1913 | Omaha, NE | 103 |
| 13. May 26, 1917 | Mattoon, IL | 101 |
| 14. June 23, 1944 | Shinnston, WV | 100 |
| 15. April 18, 1880 | Marshfield, MO | 99 |
| 16. June 1, 1903 | Gainesville & Holland, GA | 98 |
| 17. May 9, 1927 | Poplar Bluff, MO | 98 |
| 18. May 10, 1905 | Snyder, OK | 97 |
| 19. April 24, 1908 | Natchez, MS | 91 |
| 20. June 9, 1953 | Worcester, MA | 90 |
| 21. April 20, 1920 | Starkville, MI; Waco, AL | 88 |
| 22. June 28, 1924 | Lorain & Sandusky, OH | 85 |
| 23. May 25, 1955 | Udall, KS | 80 |
| 24. Sept. 29, 1927 | St. Louis, MO | 79 |
| 25. March 27, 1890 | Louisville, KY | 76 |

*Source:* National Storm Prediction Center, NOAA

same wind-resistant construction techniques used effectively in high-risk hurricane areas are being incorporated into building renovation and construction in tornado-prone areas.

## Wildfires

Wildland fires are classified into three categories: (1) a *surface fire* is the most common type and burns along the floor of a forest, moving slowly and killing or damaging trees; (2) a *ground fire* is usually started by lightning and burns on or below the forest floor; and (3) a *crown fire* spreads rapidly by wind and moves quickly by jumping along the tops of trees. Wildland fires are usually signaled by dense smoke that fills the area for miles around.

As residential areas expand into relatively untouched wildlands, people living in these communities are increasingly threatened by forest fires. Protecting structures in the wildland from fire poses special problems and can stretch firefighting resources to the limit. If heavy rains follow a fire, other natural disasters can occur, including landslides, mudflows, and floods. Once ground cover has been burned away, little is left to hold soil in place on steep slopes and hillsides. A major wildland fire can leave a large amount of scorched and barren land. These areas may not return to prefire conditions for decades. If the wildland fire destroyed the ground cover, then erosion becomes one of several potential problems.

Types of wildland fire include the following:

- *Wildland fires*. Fueled almost exclusively by natural vegetation, they typically occur in national forests and parks, where federal agencies are responsible for fire management and suppression.
- *Interface or intermix fires*. Urban/wildland fires in which vegetation and the built environment provide fuel.
- *Firestorms*. Events of such extreme intensity that effective suppression is virtually impossible, firestorms occur during extreme weather and generally burn until conditions change or the available fuel is exhausted.
- *Prescribed fires and prescribed natural fires*. Fires that are intentionally set or selected natural fires that are allowed to burn for beneficial purposes.

Severe drought conditions and the buildup of large quantities of dead trees and vegetation on the forest floors have recently led to a significant increase in wildfires in the United States. In the summer of 2002, several major wildfires raged across the country, principally in the western states. These fires consumed approximately 6 million acres of forestland, and 20 firefighters lost their lives fighting these fires.

## Landslides

Landslides occur when masses of rock, earth, or debris move down a slope. Landslides may be very small or very large, and they can move at slow to very high speeds. Many landslides have been occurring over the same terrain since prehistoric times. They are activated by storms and fires and by human modification of the land. New landslides occur as a result of rainstorms, earthquakes, volcanic eruptions, and various human activities.

Mudflows (or debris flows) are rivers of rock, earth, and other debris saturated with water. They develop when water rapidly accumulates in the ground, such as during heavy rainfall or rapid snowmelt, changing the earth into a flowing river of mud or "slurry." A slurry can flow rapidly down slopes or through channels and can strike with little or no warning at avalanche speeds. A slurry can travel several miles from its source, growing in size as it picks up trees, cars, and other materials along the way.

Lateral spreads are large elements of distributed, lateral displacement of materials. They occur in rock, but they can also occur in fine-grained, sensitive soils such as quick clays. Loose granular soils commonly produce lateral spreads through liquefaction. Liquefaction can occur spontaneously, presumably because of changes in pore-water pressures or in response to vibrations such as those produced by strong earthquakes.

Falls occur when masses of rock or other material detach from a steep slope or cliff and descend by freefall, rolling, or bouncing. Topples consist of the forward rotation of rocks or other materials about a pivot point on a hill slope.

## Tsunamis

A tsunami is a series of waves generated by an undersea disturbance such as an earthquake. From the area of the disturbance, the waves will travel outward in all directions, much like the ripples caused by throwing a rock into a pond. As the waves approach the shallow coastal waters, they appear normal and the speed decreases. Then as the tsunami nears the coastline, it may grow to great height and smash into the shore, causing much destruction.

Areas at greatest risk are less than 50 feet above sea level and within one mile of the shoreline. Tsunamis arrive as a series of successive "crests" (high water levels) and "troughs" (low water levels). These successive crests and troughs can occur anywhere from 5 to 90 minutes apart. They usually occur 10 to 45 minutes apart. The wave speed in the open ocean will average 450 miles per hour. Tsunamis reaching heights of more than 100 feet have been recorded. Most deaths during a tsunami are a result of drowning. Associated risks include flooding, polluted water supplies, and damaged gas lines.

## Volcanic Eruptions

A volcano is a mountain that opens downward to a reservoir of molten rock below the surface of the earth. Unlike most mountains, which are pushed up from below, volcanoes are built up by an accumulation of their own eruptive products—lava, ash flows, and airborne ash and dust. When pressure from gases and the molten rock becomes strong enough to cause an explosion, eruptions occur. Gases and rock shoot up through the opening and spill over, or fill the air with lava fragments. Volcanic products are used as building or road-building materials, as abrasive and cleaning agents, and as raw materials for many chemical and industrial uses. Lava ash makes soil rich in mineral nutrients.

Volcanic ash can affect people hundreds of miles away from the cone of a volcano. Several of the deaths from the Mount St. Helens volcano in 1980 were attributed to inhalation of ash. Volcanic ash can contaminate water supplies, cause electrical storms, and collapse roofs. An erupting volcano can also trigger tsunamis, flash floods, earthquakes, rock falls, and mudflows.

Sideways-directed volcanic explosions, known as "lateral blasts," can shoot large pieces of rock at very high speeds for several miles. These explosions can kill by impact, burial, or heat. They have been known to knock down entire forests. Most deaths attributed to the Mount St. Helens volcano were a result of lateral blast and trees that were blown down.

## Severe Winter Storms

Severe winter storms consist of extreme cold and heavy concentrations of snowfall or ice. A blizzard combines heavy snowfall, high winds, extreme cold, and ice storms. In the United States, the origins of the weather patterns are from four sources:

- In the Northwestern states, cyclonic weather systems from the North Pacific Ocean or the Aleutian Island region sweep massive low-pressure systems with heavy snow and blizzards.
- In the Midwestern and Upper Plains states, Canadian and Arctic cold fronts push ice and snow deep into the interior region and, in some instances, all the way down to Florida.
- In the Northeast, lake-effect snowstorms develop from the passage of cold air over the relatively warm surfaces of the Great Lakes, causing heavy snowfall and blizzard conditions.
- The Eastern and Northeastern states are affected by extra-tropical cyclonic weather systems in the Atlantic Ocean and Gulf of Mexico that produce snow, ice storms, and occasional blizzards.

## Droughts

*Drought* is defined as a water shortage caused by a deficiency of rainfall and differs from other natural hazards in three ways: (1) A drought's onset and end are difficult to determine because the effects accumulate slowly and may linger even after the apparent termination of an episode; (2) the absence of a precise and universally accepted definition adds to the confusion about whether a drought exists, and if it does, the degree of severity; and (3) drought effects are less obvious and spread over a larger geographic area.

## Extreme Heat

*Extreme heat* is defined as temperatures that hover 10 degrees or more above the average high temperature for the region and last for several weeks. Humid or muggy conditions, which add to the discomfort of high temperatures, occur when a "dome" of high atmospheric pressure traps hazy, damp air near the ground. Excessively dry and hot conditions can provoke dust storms and low visibility. Droughts occur when a long period passes without substantial rainfall. A heat wave combined with a drought is a very dangerous situation.

## Coastal Erosion

Coastal erosion is measured as the rate of change in the position or horizontal displacement of a shoreline over a period of time. It is generally associated with storm surges, hurricanes, windstorms, and flooding hazards, and may be exacerbated by human activities such as boat wakes, shoreline hardening, and dredging.

## Thunderstorms

Thunderstorms can bring heavy rains (which can cause flash flooding), strong winds, hail, lightning and tornadoes. Thunderstorms are generated by atmospheric imbalance and turbulence caused by the combination of conditions: (1) unstable warm air rising rapidly into the atmosphere, (2) sufficient moisture to form clouds and rain, and (3) upward lift of air currents caused by colliding weather fronts (cold and warm), sea breezes, or mountains.

Thunderstorms may occur singly, in clusters, or in lines. Thus it is possible for several thunderstorms to affect one location in the course of a few hours. Some of the most severe weather occurs when a single thunderstorm affects one location for an extended period. Lightning is a major threat during a thunderstorm. In the United States, between 75 and 100 Americans are hit and killed by lightning each year. A thunderstorm is classified as severe if its winds reach or exceed 58 mph, it produces a tornado, or it drops surface hail at least 0.75 inch in diameter.

Significant airplane disasters often are associated with thunderstorms and lightning. It is a myth that lightning never strikes twice in the same place. In fact, lightning will strike several times in the same place in the course of one discharge. A bolt of lightning reaches a temperature approaching 50,000 degrees Fahrenheit in a split second.

## Hailstorms

Hailstorms are an outgrowth of a severe thunderstorm in which balls or irregularly shaped lumps of ice greater than 0.75 inch in diameter fall with rain. Hailstorms occur more frequently during late spring and early summer, when the jet stream migrates northward across the Great Plains. Hailstorms cause nearly $1 billion in property and crop damage annually.

## Snow Avalanches

A snow avalanche is sliding snow or an ice mass that moves at high velocities. It can sheer trees, completely cover entire communities and highway routes, and level buildings. Natural and human-induced snow avalanches most often result from structural weaknesses within the snowpack. The potential for a snow avalanche increases with significant temperature influences.

The primary threat is loss of life of backcountry skiers, climbers, and snowmobilers as a result of suffocation when buried in an avalanche. Around 10,000 avalanches are reported each year. Since 1790, an average of 144 persons have been trapped in avalanches annually: on average, 14 injured and 14 died. The estimated annual average damage to structures is $500,000.

## Land Subsidence

Land subsidence is the loss of surface elevation caused by the removal of subsurface support; it ranges from broad, regional lowering of the land surface to localized collapse. The primary cause of most subsidence is human activities: underground mining of coal, groundwater or petroleum withdrawal, and drainage of organic soils.

**Table 2-5** Top Ten Major Disasters Ranked by FEMA Relief Costs, 1989–1999

| Event (Location, Year) | FEMA Funding* |
|---|---|
| 1. Northridge Earthquake (CA, 1994) | $6.952 billion |
| 2. Hurricane Georges (AL, FL, LA, MS, PR, USVI, 1998) | $2.394 billion |
| 3. Hurricane Andrew (FL, LA, 1992) | $1.847 billion |
| 4. Hurricane Hugo (NC, SC, PR, VI, 1989) | $1.314 billion |
| 5. Midwest Floods (IL, IA, KS, MN, MO, NE, ND, SD, WI, 1993) | $1.144 billion |
| 6. Hurricane Floyd (CT, DE, FL, ME, MD, NH, NJ, NY, NC, PA, SC, VT, VA, 1999) | $880.4 million |
| 7. Loma Prieta Earthquake (CA, 1989) | $869.0 million |
| 8. Red River Valley Floods (MN, ND, SD, 1997) | $725.1 million |
| 9. Hurricane Fran (MD, NC, PA, SC, VA, WVA, 1996) | $630.2 million |
| 10. Tropical Storm Alberto (AL, FL, GA, 1994) | $542.8 million |

*Amount obligated from the President's Disaster Relief Fund for FEMA's assistance programs, hazard mitigation grants, federal mission assignments, contractual services, and administrative costs as of July 31, 2000. Figures do not include funding provided by other participating federal agencies, such as the disaster loan programs of the Small Business Administration and the Agriculture Department's Farm Service Agency.
*Source:* FEMA, www.fema.gov

The average annual damage from all types of subsidence is conservatively estimated to be at least $125 million.

### Expansive Soils

Soils and soft rock that tend to swell or shrink because of changes in moisture content are commonly known as expansive soils. Changes in soil volume present a hazard primarily to structures that are built on top of expansive soils. The most extensive damage occurs to highways and streets. Two major groups of rocks that are prone to expansiveness and that occur more commonly in the West than East are aluminum silicate minerals (i.e., ash, glass, and rocks of volcanic origin) and sedimentary rock (i.e., clay minerals, shale).

### Dam Failures

Dam failures are potentially the worst flood events. A dam failure is usually the result of neglect, poor design, or structural damage caused by a major event such as an earthquake. When a dam fails, a gigantic quantity of water is suddenly let loose downstream, destroying anything in its path.

## TECHNOLOGICAL HAZARDS

### Fires

Fires can be triggered or exacerbated by lightning, high winds, earthquakes, volcanoes, and floods. Lightning is the most significant natural contributor to fires

**Table 2-6** U.S. Fire Losses, 1991–2000

| Year | Fires | Deaths | Injuries | Losses in Millions |
|---|---|---|---|---|
| 1991 | 2,041,500 | 4,465 | 29,375 | $10,906 |
| 1992 | 1,964,500 | 4,730 | 28,700 | $9,276 |
| 1993 | 1,952,500 | 4,635 | 30,475 | $9,279 |
| 1994 | 2,054,500 | 4,275 | 27,250 | $8,630 |
| 1995 | 1,965,500 | 4,585 | 25,775 | $9,182 |
| 1996 | 1,975,000 | 4,990 | 25,550 | $9,406 |
| 1997 | 1,795,000 | 4,050 | 23,750 | $8,525 |
| 1998 | 1,755,000 | 4,035 | 23,100 | $8,629 |
| 1999 | 1,823,000 | 3,570 | 21,875 | $10,024 |
| 2000 | 1,708,000 | 4,045 | 22,350 | $11,207 |

*Source:* National Fire Protection Association, 2000, Fire Loss in the U.S

affecting the built environment. Buildings with rooftop storage tanks for flammable liquids are particularly susceptible.

## Hazardous Materials Incidents

Hazardous materials are chemical substances, which if released or misused can pose a threat to the environment or health. These chemicals are used in industry, agriculture, medicine, research, and consumer goods. Hazardous materials come in the form of explosives, flammable and combustible substances, poisons, and radioactive materials. These substances are most often released as a result of transportation accidents or because of chemical accidents in plants.

Hazardous materials in various forms can cause death, serious injury, long-lasting health effects, and damage to buildings, homes, and other property. Many products containing hazardous chemicals are routinely used and stored in homes. These products are also shipped daily on the nation's highways, railroads, waterways, and pipelines. Varying quantities of hazardous materials are manufactured, used, or stored at an estimated 4.5 million facilities in the United States—from major industrial plants to local dry cleaning establishments or gardening supply stores.

## Nuclear Accidents

The potential danger from an accident at a nuclear power plant is exposure to radiation. This exposure could come from the release of radioactive material from the plant into the environment, usually characterized by a plume (cloudlike) formation. The area that the radioactive release may affect is determined by the amount released from the plant, wind direction and speed, and weather conditions (e.g., rain, snow) that would quickly drive the radioactive material to the ground, hence causing increased deposition of radio nuclides. Radioactive materials are composed of atoms that are unstable. An unstable atom gives off its excess energy until it becomes stable. The energy emitted is radiation. The process by which an atom changes from an unstable state to a more stable state by emitting radiation is called *radioactive decay* or *radioactivity*.

Since 1980, each utility that owns a commercial nuclear power plant in the United States has been required to have both an on-site and off-site emergency response plan as a condition of obtaining and maintaining a license to operate that plant. On-site emergency response plans are approved by the Nuclear Regulatory Commission (NRC). Off-site plans (which are closely coordinated with the utility's on-site emergency response plan) are evaluated by FEMA and provided to the NRC, who must consider the FEMA findings when issuing or maintaining a license.

Radioactive materials, if handled improperly, or radiation that is accidentally released into the environment can be dangerous because of the harmful effects of certain types of radiation on the body. The longer a person is exposed to radiation and the closer the person is to the radiation, the greater the risk. Although radiation cannot be detected by the senses (e.g., sight, smell), it is easily detected by scientists with sophisticated instruments that can detect even the smallest levels of radiation.

## Terrorism

Terrorism is the use of force or violence against persons or property in violation of the criminal laws of the United States for purposes of intimidation, coercion, or ransom. Terrorists often use threats to create fear among the public, to try to convince citizens that their government is powerless to prevent terrorism, and to get immediate publicity for their causes.

Before the September 11, 2001, attacks on New York and the Pentagon, most terrorist incidents in the United States were bombing attacks, involving detonated and undetonated explosive devices, tear gas, and pipe and fire bombs. The effects of terrorism can vary significantly from loss of life and injuries to property damage and disruptions in services such as electricity, water supply, public transportation, and communications.

One way governments attempt to reduce people's vulnerability to terrorist incidents is by increasing security at airports and other public facilities. The U.S. government also works with other countries to limit the sources of support for terrorism. The Federal Bureau of Investigation (FBI) categorizes terrorism in the United States as one of two types: domestic terrorism or international terrorism. Domestic terrorism involves groups or individuals whose terrorism activities are directed at elements of government or population without foreign direction. International terrorism involves groups or individuals whose terrorist activities are foreign-based and/or directed by countries or groups outside the United States or whose activities transcend national boundaries.

## Biological and Chemical Weapons

Biological agents are infectious microbes or toxins that are used to produce illness or death in people, animals, or plants. Biological agents can be dispersed as aerosols or airborne particles. Terrorists may use biological agents to contaminate food or water because they are extremely difficult to detect. Chemical agents kill or incapacitate people, destroy livestock, or ravage crops. Some chemical agents are odorless and tasteless and are difficult to detect. They can have an immediate effect (a few seconds to a few minutes) or a delayed effect (several hours to several days).

# RISK ASSESSMENT

Most practitioners and academics refer to the term *risk assessment* as a process or methodology that can be used for evaluating risk. In this context, risk is defined as (1) the probability and frequency of a hazard occurring, (2) the level of exposure of people and property to the hazard, and (3) the effects or costs, both direct and indirect, of this exposure. There are various approaches to developing a risk assessment methodology, ranging from qualitative to quantitative, as well as several computer-based models for natural hazard risk assessment, currently in use in the United States and Japan.

The validity and use of any risk assessment are determined by the quality and availability of data. Because these two factors are still unknown and will not be determined until the in-country risk templates have been compiled, the determination of the most effective approach will not be made until the data has been collected and reviewed; however, a general discussion of the suggested approach will be undertaken.

As mentioned previously, various accepted methodologies could be applied. These include the risk matrix approach that is qualitative and is designed to support risk management planning and decision making. The Composite Exposure Indicator (CEI) approach is based on the effects of a single or multiple hazards on a series of indicator variables focused primarily on infrastructure, such as roads, pipelines, hospitals, public water supply, and so on. The CEI is a measure of exposure of 14 variables that produces a number that is then correlated to the population affected. Numerous approaches result in vulnerability analyses that have been applied to earthquake and hurricane (coastal) hazards. The differences between these approaches often relate to *how* direct costs or *if* indirect costs are measured.

Common to most of these methodologies is a series of essential elements or steps that must be undertaken. In general these steps are as follows:

1. *Identify and characterize the hazard.* What are the characteristics of the hazard (e.g., high-velocity winds, ground shaking)? What causes the hazard event, and how does it trigger or relate to other hazards?
2. *Evaluate each hazard for the severity and frequency.* What is the probability of a hazard event happening annually, every 10 years, once a century? What factors enhance or deter the probabilities? What measurements or scales can be applied to determine severity? Could other factors influence severity and frequency (e.g., El Niño, global warming)?
3. *Estimate the risk.* Identify and quantify what will be affected by the hazard event. This step imposes the human and built environment that could be affected, damaged, or disrupted by a hazard event. Included in the analysis would be the general building stock (commercial and residential), inventories of lifelines, and essential, critical facilities. Population and development concentrations would be included.

4. *Determine the potential societal and economic (direct) effects and the indirect effects or costs.* In estimating direct economic losses, data that would be included is the cost of repair or replacement of damaged structures or lifelines, nonstructural damage, loss of contents and business inventory, and related loss of function costs. Agricultural (crop) losses figure prominently in this category. Other costs could be income loss, relocation costs, and rental losses that occur as a consequence of the event.

Social costs are predominantly categorized as casualties, injuries, displaced households, and the cost of sheltering. Indirect effects and costs are more difficult to calculate and the data more difficult to obtain. Examples of indirect economic effects can include increase in unemployment, business interruption and loss of production, reduction in demand and consumer spending, and tax base losses. Indirect losses are more easily calculated at the local and regional levels because the information needed relative to population, employment, and tax base and the nature of the economy and businesses is more easily identified.

The costs to federal, state, and local governments, individuals, and businesses of responding to disaster events are often not incorporated into the cost effect equation, but in many cases these costs have a significant effect on agencies' budgets and should be considered.

Two other steps should be included in looking at a risk assessment methodology:

5. *Determine the acceptable level of risk.* An analysis is undertaken of the information or data assembled in steps 1 to 4 to establish an acceptable level of risk. This means simply: What level of damage or impact will be tolerated? Societal effects and the less tangible, direct and indirect costs make this evaluation a more difficult part of the process. Compounding this difficulty are the public perception of risk and the political consequences of taking or not taking action to address the risks.
6. *Identify risk-reduction opportunities.* This critical step takes the risk assessment methodology beyond process to decision making and action. At this point, cost-effective actions that will reduce or mitigate unacceptable risks should be identified and implemented. A variety of structural and nonstructural alternatives can be combined with technology, legislation, and other solutions to design a risk-reduction implementation plan consistent with the degree of risks.

## TECHNOLOGY

The nation's ability to identify hazards and quantify risk has significantly improved in the last 10 years. Technological advances have refined the ability to identify and understand the nature of hazards and develop better risk assessment methods. Recent technological advances include the use of satellite imagery and radar to map ever-changing floodplains and areas of coastal erosions, the FEMA-developed HAZUS loss estimation model that provides us with loss estimates from various earthquake scenarios, and the technology that has created safe rooms for homes in tornado-prone areas. The research and scientific agencies of the federal government and the university community continue to develop new approaches to measuring, mapping, and predicting natural hazards. With the reality of September 11, technology is focusing on new methods to detect, prevent, or provide an antidote for the various biological and chemical agents that could be used in a terrorist event.

## CONCLUSION

With increased knowledge comes increased responsibility. Providing states and communities with hazard information and these tools requires that they take action to address the hazards and risks. Emergency management provides the impetus for incorporating these considerations into planning and governing communities.

Hazards will continue to exist. Some hazards, particularly those that are technological, may be reduced by collective efforts, but the ability to control or eliminate natural hazards is questionable. Recent efforts to undo some of the channelization and flood control projects undertaken by the U.S. Army Corps of Engineers are vivid examples of the inability to control nature; however, there is still a strong argument for an increased emphasis on improved science in hazard identification and increased financial support for hazards mapping.

As knowledge increases, the economic and social logic of applying long-term solutions for reducing the risks posed by these hazards through mitigation will gain momentum. The emergency management profession should be providing leadership to take advantage of this momentum.

# 3. The Disciplines of Emergency Management: Mitigation

## INTRODUCTION

Disasters are a reality of living in the natural world. Despite humans' attempts to control nature, dating back to the early Egyptians and continuing to this century's massive flood control efforts, natural hazards continue.

Over the last decade, the social and economic costs of disasters to the United States and throughout the world have grown significantly. From the period 1990 to 1999, FEMA spent more than $25.4 billion to provide disaster assistance in the United States. During the 1990s, the economic toll of natural disasters topped $608 billion worldwide, more than the previous four decades combined. The causes of this growth are myriad. Climatological changes such as El Niño, global warming, and sea level rise are one factor. Add to these changes the effects of societal actions such as increased development, deforestation and clear-cutting, migration of population to coastal areas, and filling in of floodplains, and a recipe for disaster results.

The discipline of mitigation provides the means for reducing these impacts. *Mitigation* is defined as a sustained action to reduce or eliminate risk to people and property from hazards and their effects. This discussion of mitigation focuses on natural hazards mitigation efforts and programs in the United States. Techniques for mitigation of technological hazards will be referenced, but the body of knowledge and applications in this area are still evolving; however, many of the successful natural hazards techniques such as building codes do have applicability to technological hazards.

The function of mitigation differs from the other emergency management disciplines because it looks at long-term solutions to reducing risk as opposed to preparedness for hazards, the immediate response to a hazard, or the short-term recovery from a hazard event. Mitigation is usually not considered part of the emergency phase of a disaster as in response or as part of emergency planning as in preparedness. The definition lines do get a little blurred regarding recovery. As discussed in Chapter 5, applying mitigation strategies should be a part of recovery from disaster; however, even in this context, these are actions that will reduce the impacts, or risks, over time.

The recovery function of emergency management still represents one of the best opportunities for mitigation, and until recently, this phase in a disaster plan provided the most substantial funding for mitigation activities. Recently there has been a trend toward greater federal spending on predisaster mitigation, which is discussed later in this chapter.

Another difference sets mitigation apart from the other disciplines of emergency management. Implementing mitigation programs and activities requires the participation and support of a broad spectrum of players outside of the traditional emergency management circle. Mitigation involves, among others, land-use planners, construction and building officials, both public and private, business owners, insurance companies, community leaders, and politicians.

The skills and tools for accomplishing mitigation (i.e., planning expertise, political acumen, marketing and public relations, and consensus building) are different from the operational, first-responder skills that more often characterize emergency management professionals. In fact, historically, emergency management professionals have been reluctant to take a lead role in promoting mitigation. A state director of emergency management once said words to the effect: "I will never lose my job for failing to do mitigation, but I could lose my job if I mess up a response."

With the exception of the fire community, who were early leaders in the effort to mitigate fire risks through support for building codes, code enforcement, and public education, the emergency management community has remained focused on their response and recovery obligations; however, this trend is changing for several reasons. Leadership at the federal level, larger disasters, substantial increases in funding, and more value and professionalism in emergency management have resulted in greater acknowledgement of the importance of mitigation.

This chapter discusses the tools of mitigation, the impediments to mitigation, federal programs that support mitigation, and several case studies that demonstrate how these tools have been applied to successfully reduce various risks.

## TOOLS FOR MITIGATION

Over the years, the United States has made great strides in reducing the number of deaths that occur in natural disasters. Through building codes, warning systems, and public education, the number of deaths and casualties from natural disasters in the last century has significantly declined; however, economic effects and property damages have escalated. Many people believe that these costs are preventable and that the tools exist to dramatically reduce these costs.

Technological disasters such as the Oklahoma City bombing and the terrorist attacks of September 11, 2001, are not as easy to analyze. There is much speculation about how improved intelligence and security could reduce the human effects of these disasters. From a property perspective, many people believe that some reduction in impacts could be achieved through application of traditional mitigation techniques such as improved building construction for blast effects. Other technological disasters such as the Valdez oil spill, the Three Mile Island emergency, and so on could have been prevented through better inspections, training, education, and exercises. These measures reflect good preparedness activities more than mitigation. In any case, further research and analyses are needed to answer the questions posed by the effects of terrorist events and similar technological hazards.

Most practitioners agree that the primary intent of mitigation is to ensure that fewer communities and individuals become victims of disasters. The goal of mitigation is to create economically secure, socially stable, better built, and more environmentally sound communities that are out of harm's way.

The following widely accepted mitigation tools are used to reduce risk:

- Hazard identification and mapping
- Design and construction applications
- Land-use planning
- Financial incentives
- Insurance
- Structural controls

## Hazard Identification and Mapping

This is the most obvious tool for mitigation. You can't mitigate a hazard if you don't know what it is or whom it affects. The most essential part of any mitigation strategy or plan is an analysis of what the hazards are in a particular area. The resources for hazards identification are numerous. The federal government has extensive programs that map virtually every hazard, and these products are available to communities. FEMA's National Flood Insurance Program (NFIP) provides detailed flood maps and studies, and the U.S. Geological Survey (USGS) provides extensive earthquake and landslide studies and maps. Many state agencies have refined the products for hazards identification. For example, special soil stability studies and geological investigations, which are required in some parts of California, further refine this analysis.

Geographic information systems (GIS) have become ubiquitous and staples for all local planning organizations. What is often missing from the available tools is the ability to superimpose the human and built environment onto the hazards, thereby providing a quantified level of risk. FEMA has developed one such tool called HAZUS. HAZUS is a nationally applicable methodology for estimating losses from earthquakes at the community or regional level. FEMA is currently expanding HAZUS to cover hurricane or wind losses and floods.

## Design and Construction Applications

The design and construction process provides one of the most cost-effective means of addressing risk. This process is governed by building codes, architecture and design criteria, and soils and landscaping considerations. Code criteria that support risk reduction usually apply only to new construction, substantial renovation, or renovation to change the type or use of the building. Enactment of building codes is the responsibility of the states, and most state codes are derivatives of one of the three model codes, which reflect geographical differences across the United States. Some states delegate code adoption responsibility to more local governmental authorities. Because of cost, codes that require rehabilitation of existing potentially hazardous structures have been rarely implemented. The Los Angeles seismic retrofit ordinance is a rare example. The case study of the Virgin Islands at the end of this chapter illustrates the importance of building codes to mitigation.

The construction process offers other opportunities. For example, using fire-retardant building materials such as slate instead of wood for roofing is important in areas of wildland/urban interface such as Oakland, California. Constructing houses on pilings allows for uninterrupted flow of high-velocity waves in coastal areas.

Landscaping is particularly critical in areas of potential wildfires because vegetation close to structures can become fuel for a fire. Clearing, grading, and siting all have potential impacts to soil stability and erosion and can be included as part of a design or building permit review process.

The federal government has made a significant investment in developing technical guidance for improving the building and construction of structures in hazard areas, particularly earthquake, wind, and flood-prone areas. There has been some discussion of developing a National Code to support mitigation efforts. Because the Constitutional responsibility for public health and safety resides with the states, a National Code developed by the federal government is not politically feasible or practical.

## Land-Use Planning

Mitigation programs are most successful when undertaken at the local level, where most decisions about development are made. The strategies for land-use planning offer many options for effecting mitigation, including acquisition, easements, storm water management, annexation, environmental review, and floodplain management plans. It also encompasses a myriad of zoning options such as density controls, special uses permits, historic preservation, coastal zone management, and subdivision controls.

Land-use planning was one of the earliest tools used to encourage mitigation. In 1968 Congress passed the National Flood Insurance Act that established the NFIP. This act required local governments to pass a floodplain management ordinance in return for federally backed, low-cost flood insurance being available to the community. This act started one of the largest federal mapping efforts because the government promised local governments that they would provide them with the technical tools to determine where the floodplains were in their communities so they could steer development away from these areas. A more complete discussion of the NFIP can be found later in this chapter.

Moving structures out of harm's way through property acquisition is clearly the most effective land-use planning tool, but it is also the most costly. Following the Midwest floods of 1993, FEMA worked with Congress to make property acquisition more feasible by providing a substantial increase in funding for acquisition after a disaster. The case study on Missouri at the end of this chapter provides documentation on how well an acquisition strategy can work.

There are many other examples of how land-use planning and ordinances can promote risk reduction. The North Carolina coastal setback ordinance seeks to preserve the fragile and eroding coastlines of its barrier islands. The Alquist-Priola Act in California limits development near known earthquake faults.

## Financial Incentives

This is one of the emerging areas for promoting mitigation. Among the approaches being used by localities to reduce risk are creation of special tax assessments, passage of tax increases or bonds to pay for mitigation, relocation assistance, and targeting of federal community development or renewal grant funds for mitigation.

The economic effects of repetitive flooding led the citizens of Napa, California, and Tulsa, Oklahoma, to pass small tax increases to pay for flood-mitigation activities. In both cases, the tax had minimal effect on the community citizens but had a major effect in reducing the potential economic losses from future floods. Berkeley, California, has passed more than 10 different bond issues to support seismic retrofit of public buildings, schools, and private residences.

Funding from the Community Development Block Grant (CDBG), a HUD program, has been used extensively to support local efforts at property acquisition and relocation. These funds have been used to meet the nonfederal match on other federal funding, which has often been a stumbling block to local mitigation. Other federal programs of the Small Business Administration (SBA) and the Economic Development Administration provide financial incentives for mitigation.

Other emerging areas of financial tools include special assessment districts, impact fees, and transfer of development rights. All of these tools provide either incentives or penalties to developers as a means of promoting good risk-reduction development practices.

## Insurance

Some people would argue with the inclusion of insurance as a mitigation tool. Their reasoning is that insurance by itself really only provides for a transfer of the risk from the individual or community to the insurance company. Although this is true, the National Flood Insurance Program (NFIP) is the prime example of how, if properly designed, the insurance mechanism can be a tool for mitigation. The NFIP is considered to be one of the most successful mitigation programs ever created.

The NFIP was created by Congress in response to the damages from multiple, severe hurricanes and inland flooding and the rising costs of disaster assistance after these floods. At that time, flood insurance was not readily available or affordable through the private insurance market. Because many of the people being affected by this flooding were low-income residents, Congress agreed to subsidize the cost of the insurance so the premiums would be affordable. The idea was to reduce the costs to the government of disaster assistance through insurance. The designers of this program, with great insight, thought the government should get something for their subsidy. So in exchange for the low-cost insurance, they required that communities pass an ordinance directing future development away from the floodplain.

The NFIP was designed as a voluntary program and, as such, did not prosper during its early years, even though flooding disaster continued. Then in 1973, after Hurricane Agnes, the legislation was modified significantly. The purchase of federal flood insurance became mandatory on all federally backed loans. In other words, anyone buying a property with a Veterans Administration (VA) or Federal Housing Administration (FHA) loan had to purchase the insurance. Citizen pressure to buy the insurance caused communities to pass ordinances and join the NFIP. The NFIP helped the communities by providing them with a variety of flood hazard maps to define their flood boundaries and set insurance rates.

The 1993 Midwest floods triggered another major reform to the NFIP. This act strengthened the compliance procedures. It told communities that if they didn't join the program, they would only be eligible for disaster assistance one time. Any further request would be denied. As a positive incentive, the act established a Flood Mitigation Assistance (FMA) fund for flood planning, flood mitigation grants, and additional policy coverage for meeting the tougher compliance requirements such as building elevation.

Over the years, the NFIP has created other incentive programs such as the Community Rating System. This program rewards those communities that go beyond the minimum floodplain ordinance requirements with reduced insurance premiums. The NFIP represents one of the best public/private partnerships. Through the Write Your Own program, private insurers are given incentives to market and sell flood insurance.

Today more than 20,000 communities in the NFIP have mitigation programs in place. Other attempts have been made to duplicate this program for wind and earthquake hazards, but these have not received the support necessary to pass in the Congress. If another major earthquake occurs, the issue of creating a federally supported earthquake or all-hazards insurance will resurface.

## Structural Controls

Structural controls are controversial as a mitigation tool. Structural controls have usually been used to protect existing development. In doing so, they can have both positive and negative effects on the areas they are not protecting. In addition, as the name implies, they are used to control the hazard, not reduce it. Invariably, as was seen so graphically in the Midwest floods, the structures lose control and nature wins; however, in some circumstances, structural controls are the only alternative.

The most common form of structural control is the levee. The U.S. Army Corps of Engineers has designed and built levees as flood control structures across the United States. Levees are part of the aging infrastructure of America. As mitigation tools, they have obvious limitations. They can be overtopped or breached, as in the 1993 Midwest floods, they give residents a false sense of safety that often promotes increased development, and they can exacerbate the hazard in other locations. After the 1993 floods, a major rethinking of dependency on levees has occurred. Efforts are being made to acquire structures built behind the levees, new design criteria are being considered, and other more wetland-friendly policies are being adopted. For a city like New Orleans, however, which is built below sea level and where relocation is impractical, levees can be used effectively to protect flood-prone areas.

Other structural controls are intended to protect along coastal areas. Seawalls, bulkheads, breakwaters, groins, and jetties are intended to stabilize the beach or reduce the impacts of wave action. These structures are equally controversial because they protect in one place and increase the damage in another. The shore of New Jersey is a prime example of the failure of seawalls as a solution to shoreline erosion problems. Cape May, New Jersey, where cars used to be raced on the beach, lost all of its beachfront. An ongoing beach replenishment project is the only thing that has brought some of it back.

## IMPEDIMENTS TO MITIGATION

If so many tools can be applied, why haven't risk-reduction and mitigation programs been more widely applied? There are several factors, including denial of the risk, political will, costs and lack of funding, and the taking issue. Despite the best technical knowledge, historic occurrence, public education, and media attention, many individuals don't want to recognize that they or their communities are vulnerable. Recognition requires action and it could have economic consequences as businesses decide to locate elsewhere if they find the community is at risk. Some people are willing to try to beat the odds, but if a disaster strikes, they know the government will help them out. Gradually, attitudes are changing. Potential liability issues are making communities more aware, media attention to disasters has brought public pressure, and the government has provided both incentives for taking action and penalties for not taking action.

As previously mentioned, mitigation provides a long-term benefit. The U.S. political system tends to focus on short-term rewards. Developers are large players in the political process and are often concerned that mitigation means additional costs. Mitigation strategies and actions require political vision and will. As Tip O'Neill, former Speaker of the U.S. House of Representatives, said: "All politics is local." Well, so is mitigation. Local elected officials are the individuals who have to promote, market, and endorse adopting risk reduction as a goal. For many elected officials, the development pressures are too much, funding is lacking, and other priorities dominate their agendas; however, with the increasing attention to the economic, social, and political costs of not dealing with their risks, more elected officials are recognizing that they can't afford to *not* take action.

Mitigation costs money. Most mitigation of new structures or development can be passed on to the builder or buyer without much notice. Programs to retrofit existing structures or acquisition and relocation projects are expensive and almost always beyond the capacity of the local government. Funding for mitigation comes primarily from federal programs that need to be matched with state or local dollars. As state and local budgets constrict, their ability to match is reduced. Strong arguments can be made that it is in the best financial interest of the federal government to support mitigation. These arguments and a series of large disasters resulted in substantial increases in federal funding, including new monies for predisaster mitigation, but the fact remains that mitigation needs far outweigh mitigation funding.

Many mitigation actions involve privately owned property. A major legal issue surrounding this is the taking issue. The Fifth Amendment to the U.S. Constitution prohibits the taking of property without just compensation. What constitutes a taking, under what circumstances, and what is just compensation have been the focus of numerous legal cases. Several have dealt with the use of property in the floodplain and the use of oceanfront property on a barrier island. The decisions have been mixed, and taking will continue to be an issue in implementing mitigation programs and policies.

# FEDERAL MITIGATION PROGRAMS

FEMA is responsible for most of the programs of the federal government that support mitigation; this section focuses on these programs. As noted earlier, the Small Business Administration (SBA), Economic Development Administration (EDA), and HUD have policies that support mitigation. The PATH program at HUD supports incorporating mitigation into public housing. The Environmental Protection Agency (EPA) has several programs in floodplain management and in 2002 initiated a new pilot program for national watersheds. The National Earthquake Hazards Reduction Program, which is described in a following section, includes several other federal agencies; however, the predominant federal agency involved in disaster mitigation is FEMA.

FEMA's programs include the NFIP (described earlier in the chapter), the Hazard Mitigation Grant Program (HMGP), the Pre-Disaster Mitigation Program (PDM), the National Earthquake Hazard Reduction Program (NEHRP), the National Hurricane Program, the National Dam Safety Program, and the Fire Prevention and Assistance Grant Program.

## The Hazard Mitigation Grant Program

HMGP is the largest source of funding for state and local mitigation activities. This program provides grants to state and local governments to implement long-term hazard mitigation programs after a major disaster has been declared by the President. HMGP projects must reduce the risk, and the benefits of the project must exceed the costs.

Examples of activities supported by HMGP include the following:

- Acquisition of property on a voluntary basis and commitment to open use of the property
- Retrofitting of structures and lifelines
- Elevation of structures
- Vegetation management programs
- Building code enforcement
- Localized flood-control projects
- Public education and awareness

This program was enacted by Congress in 1988 as part of the Robert T. Stafford Act that was a major reworking of federal disaster policy. Besides creating the HMGP, it established a cost sharing of disaster assistance by the states. At the time the formula for state HMGP funding was 15 percent of the public assistance costs, and it had a 50 percent federal, 50 percent state cost share.

From the period 1988 to 1993, many states did not take advantage of the HMGP funding because it was difficult to meet the matching requirements, even though the

15 percent cap was often not very much. After the devastation of the 1993 Midwest floods, Congressman Volkmer from Missouri championed a change to the legislation that would significantly increase the states' ability to mitigate. Congress amended the legislation to allow for a 75 percent federal, 25 percent state match, and dramatically increased the amount of funding to 15 percent of the total disaster costs. The rationale for these changes was to aggressively work to move people and structures out of the floodplain. As the Missouri case study at the end of this chapter documents, the rationale was sound.

HMGP has allowed states to hire staff to work on mitigation and requires development of a State Hazard Mitigation Plan as a condition of funding. This program brought about a change in the emergency management community at the state and local levels. With adequate funding, states and localities began to hire staff designated to work on mitigation.

HMGP has its detractors and, in 2002, the Federal Office of Management and Budget (OMB) proposed that this program be eliminated in favor of a new predisaster competitive grant program. It is unlikely that Congress will eliminate this program.

## Pre-Disaster Mitigation Program

Through the Disaster Mitigation Act of 2000, Congress approved creation of a national PDM to provide mitigation funding not dependent on a disaster declaration. The genesis of PDM was an initiative of the Clinton administration called Project Impact: Building Disaster-Resistant Communities. Project Impact grew out of the devastating disasters of the 1990s. Many of the communities hit by these disasters took months and even years to recover emotionally and financially. James Lee Witt, then Director of FEMA, questioned the wisdom of spending more than $2.5 billion per year on disaster relief and not a penny to reduce disasters before they happen. The mitigation tools and techniques were available, so why not work to prevent individuals and communities from becoming victims of disasters? With a small amount of seed money, FEMA launched Project Impact in 1997 in seven pilot communities.

The concept behind the initiative was simple. The mitigation activities had to be designed and tailored to the hazards in that community, and all sectors of the community had to become involved in order for it to be effective and sustainable. Project Impact brought the business community under the emergency management umbrella. Communities were asked to achieve the following four goals:

1. Build a community partnership.
2. Assess the risks.
3. Prioritize risk-reduction actions.
4. Build support by communicating your actions.

By 2001, more than 200 communities were participating in Project Impact, and Congress had appropriated $25 million to the initiative. Seattle, Washington, was one of the original pilot communities. In 2002, when a 6.8 earthquake struck Seattle, the mayor attributed the success of their Project Impact activities for the minimal damages and prompt recovery. The Tulsa case study provides an example of a Project Impact community.

In 2002, the Bush administration decided to drop the Project Impact name and concept in exchange for a competitive grant program as their approach to PDM. They have requested $300 million and proposed that this program replace both Project Impact and the HMGP.

## The National Earthquake Hazard Reduction Program

The goal of the NEHRP is to reduce the risks to life and property from future earthquakes in the United States through the establishment and maintenance of an effective earthquake hazards-reduction program. FEMA works as the lead organization in this program, along with the National Institute of Standards and Technology (NIST), the National Science Foundation (NSF), and the USGS. The NEHRP works to improve understanding, characterization, and prediction of hazards and vulnerabilities; improve model building codes and land-use practices; reduce risk through postearthquake investigations and education; develop and improve design and construction techniques; improve mitigation capacity; and accelerate application of research results.

The NEHRP provides funding to states to establish programs that promote public education and awareness, planning, loss estimation studies, and some minimal mitigation activities. FEMA supports state and local governments by providing HAZUS. HAZUS is a tool for communities to use for estimating potential losses from natural hazards.

## The National Hurricane Program

This FEMA program supports activities at the federal, state, and local level that focus on the physical effects of hurricanes, improved response capabilities, and new mitigation techniques for the built environment. The program has done significant work in storm surge modeling and evacuation planning, design and construction of properties in hurricane-prone areas, and public education and awareness programs for schools and communities. The amount of funding that FEMA receives for this program is in the range of $3 million annually, which is clearly not commensurate with the risk.

## The National Dam Safety Program

The National Dam Safety Program Act of 1996 formally established the National Dam Safety Program and named the Director of FEMA as its coordinator. Initiatives under the act include funding to the states to establish and maintain dam safety programs; training for state dam safety staff and inspectors; technical and archival research in dam safety; education of the public in the hazards of dam failure and related matters; the establishment of the National Dam Safety Review Board; and support for the Interagency Committee on Dam Safety. This act, which is part of the Water Resources Development Act of 1996, was authorized through 2002.

### The Fire Prevention and Assistance Act

This program was created in 2001 to address the needs of the nation's paid and volunteer fire departments and to support prevention activities. Congress had long-standing concerns about the status of this first-responder community. New threats from potential biochemical terrorism, increasing wildfire requirements, and a stagnant search-and-rescue capability provided the rationale for funding this program. This multimillion-dollar grant program provides competitive grants to fire companies throughout the United States. In the wake of the September 11 events, the appropriations for this program tripled in 2002.

### FEMA'S ASSISTANCE TO FIREFIGHTERS GRANT PROGRAM

The purpose of the program is to award one-year grants directly to fire departments of a state to enhance their abilities with respect to fire and fire-related hazards. This program seeks to identify departments that lack the basic tools and resources necessary to protect the health and safety of the public and their firefighting personnel. The primary goal is to provide assistance to meet these needs.

Assistance to Firefighters Grant Program Fiscal Year 2002 Award Recipients (through August 12, 2002)

| Categories | No. of Awards | Amount of Awards |
|---|---|---|
| Fire Operations & Firefighter Safety | 506 | $31,915,961 |
| Fire Prevention | 76 | $3,330,848 |
| Firefighting Vehicles | 158 | $19,643,175 |
| Emergency Medical Services | 4 | $57,067 |
| **Total** | **744** | **$55,463,051** |

*Source:* FEMA, www.fema.gov

## CONCLUSION

Disasters occur in every state. The direct costs of these events are staggering, but the indirect effects to the economy and the social fabric of communities is even worse. Mitigation works. The case studies included in this chapter are just a few examples of successful, sustained programs that are reducing risk and making communities safer. Mitigation programs exist at all levels of government, and there is a growing interest in the private sector for taking mitigation actions to reduce their risk exposure. To many people, even in a time when terrorism preoccupies the emergency management psyche, mitigation is—and should be—the future direction of emergency management.

# CASE STUDIES

## CASE STUDY: TULSA SAFE ROOM PROGRAM

Tulsa, Oklahoma, lies in the heart of Tornado Alley. Tornadoes with major damage have hit Tulsa on average of every four or five years. Most recently, the May 3, 1999, tornadoes killed 44 people and decimated communities throughout Oklahoma. As a result of these storms, the President declared a major disaster. Oklahoma was provided the opportunity to take advantage of new construction technology to mitigate the effects of tornadoes (see Figures 3-1 and 3-2).

The concept of "safe room" construction was developed and pilot tested in 1998 by the Wind Engineering Research Center of Texas Tech University with financial support from FEMA. Safe rooms are anchored and armored rooms that provide shelter during tornadoes, even above ground. Tulsa proposed to FEMA that it use its HMGP funding provided through the President's declaration to provide grants to homeowners to build safe rooms in their homes.

Under their Project Impact designation, Tulsa formed a coalition of partners, including FEMA, Oklahoma State Emergency Management, Home Builders of Greater Tulsa, Tulsa Public Works, State Farm Insurance, and other community partners. This coalition then agreed on building and construction standards, permitting, certification and compliance procedures, and public education and awareness programs. This coalition set as their goal to build a tornado safe room in every newly constructed and existing home by the year 2020.

**Figure 3-1** November 23, 2001, Tulsa, Oklahoma (Disaster Alley in the Eastland Mall). A safe room wall section is shown here. The insulated concrete form is cut away to show reinforcing steel. The cavity is filled with concrete. Photo by Kent Baxter/ FEMA News Photo.

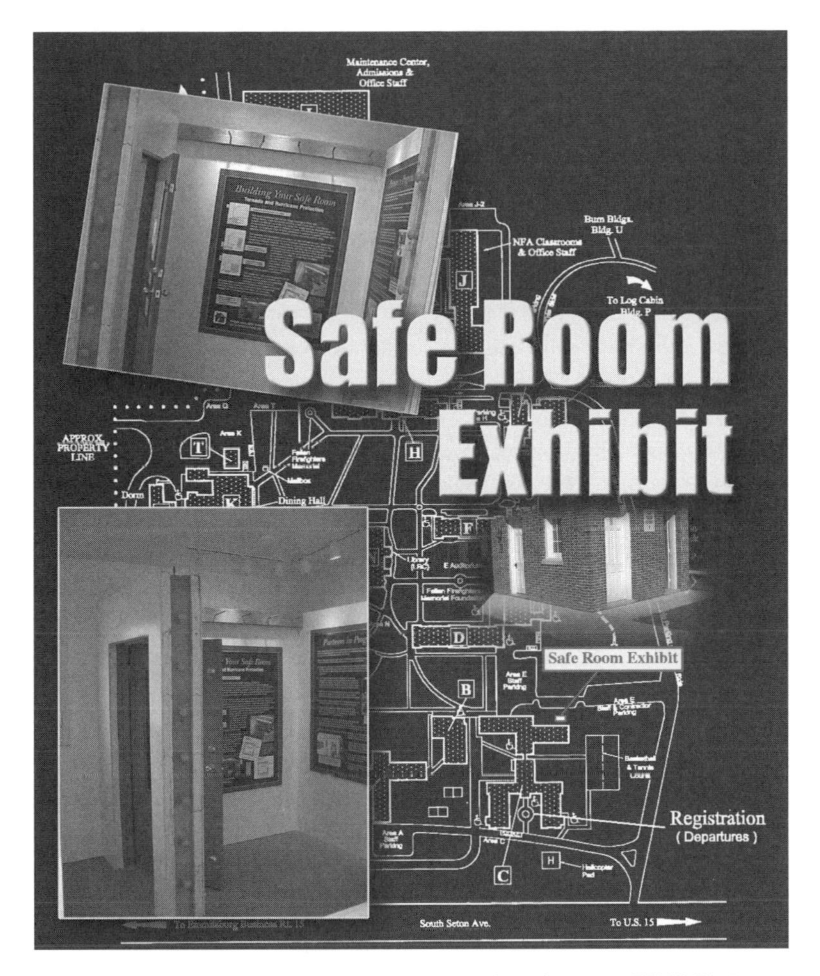


**Figure 3-2** Exhibit of techniques for a tornado safe room. FEMA Photo.

This program was supported through a variety of public and private funding, but the major key to its success was the partnership of the building and construction community.

Tulsa builders embraced the safe room concept and quickly made it a positive marketing tool for their business. The city continued to encourage growth of the program by providing certain financial incentives. Eleven major Tulsa builders launched the first safe room subdivision in a new upscale residential area of Tulsa. It is believed to be the first safe room subdivision in Oklahoma, and perhaps the first in the nation, financed entirely by private builders.

The program continues to expand not just within Tulsa and Oklahoma, but to other states and communities in "tornado alley" as well. Within Tulsa, wheelchair-accessible safe rooms have been designed and built. The next step is building safe rooms in public buildings and schools. The technology exists, but the societal questions of size, access, and quantity of space and related issues are still being worked on.

The Tulsa safe room project provides an excellent example of taking advantage of the opportunity afforded in the postdisaster climate. Its success provides an even better example of how building coalitions, particularly with the private sector, ensure sustainability of the mitigation program.

## CASE STUDY: THE CASTAIC UNION SCHOOL DISTRICT

The Castaic Union School District, located in Southern California, is a case study that demonstrates the threat from multiple hazards. After the 1994 Northridge Earthquake, Castaic Union School District conducted a study of the earthquake-related risks that threatened their elementary and middle schools and administration buildings. The assessment revealed that earthquake-related structural damage was not the only risk the school district faced.

The district maintained and operated 63 buildings (77,000 square feet of usable space) in Northern Los Angeles County that consisted of a mix of permanent and portable structures with construction dates as far back as 1917. These structures service approximately 1,200 students and 115 staff members. The San Andreas and San Gabriel fault systems, two of the most active faults in the country, pass through the area in which the district is located. In addition, the USGS has concluded that significant new earthquake activity may occur along both the San Andreas and San Gabriel systems.

These factors led the Castaic Union School District to conclude in their study that the probability of a large earthquake affecting their facilities was high. They also learned, however, that the risk went well beyond possible damages caused by ground shaking. Along with the expected seismic damage, the study revealed two additional threats: flooding from the Castaic Dam and fire or explosion from a rupture in nearby oil pipelines.

The district's risk assessment study indicated that the school buildings were located within the inundation area of the Castaic Dam (located only 1.7 miles upstream). If the dam were to fail, the school buildings and their occupants would be inundated with catastrophic flooding. The 2,200-acre reservoir above the dam could release nearly 105 billion gallons of water, inundating the area below the dam with 50 feet of water. In 1992, the California Department of Water Resources (DWR) reexamined the seismic performance of the dam. Based on the analyses, the DWR considers the dam to meet all current safety requirements and to be able to resist failure caused by the maximum credible earthquake; however, the district's risk assessment concluded the probability the Castaic Dam will fail is never zero.

Along with the threat posed by the Castaic Dam, the study also revealed that the buildings were at high risk of damage from both fire and explosion if nearby pipelines failed. Two high-pressure crude oil pipelines currently cross the campus (a 1925 gas-welded pipeline and a 1964 modern arc-welded steel pipeline), both of which could rupture during ground shaking or ground displacement in earthquakes. An analysis of the lines and the fault conditions near the district indicated a 35 percent chance of failure somewhere in the Castaic area as a result of any large earthquake.

This information caused alarm about the safety of the district's facilities. In the event of a pipeline failure, a fire or explosion could result from the ignition of the released oil, putting both facilities and people at great risk. Additionally, the ability to prevent a nearby fire from spreading would be limited by the decreased reliability of water lines and hydrants, as well as the increased demands on emergency fire services after an earthquake.

Using the results of the district's risk analysis, it was determined that the potential economic costs from either a dam failure or oil pipeline break following an earthquake were enormous. The first potential cost to the school district would be incurred from both building and content damage. Replacement of the school buildings would cost an estimated $7.7 million. Second, if such an earthquake occurred, alternate school facilities would have to be located and rented at an estimated cost of more than $500,000 per year. Third, the community would have to absorb the costs of losing the educational services provided by the district in the time period between the actual loss of the facilities and the relocation to temporary facilities. The school district calculated the cost of the lost public services based on the operating expenses required to provide the services. The daily cost of lost educational services was estimated at $28,601.

In addition to these direct and indirect financial losses, the risk of earthquake-related casualties in the district's facilities was determined to be significant. In an earthquake-induced dam failure, the predicted speed of inundation on the campus caused the risk of casualties to be very high. When calculating this risk, a casualty rate of 250 individuals was determined based on the average hourly rate of campus usage in a typical week. In the event of a dam failure during school hours, the loss of life could be as high as 1,200 students and 115 faculty members. In an earthquake-induced potential pipeline failure, the district calculated a casualty rate of 9 individuals and injury rate of 45 individuals. Once again, the actual number of casualties increases dramatically if the earthquake and pipeline failure occur during school hours.

Through the cost-benefit analysis, the district determined that the most feasible method to reduce their risks would be to condemn the structures on the old, high-risk site and relocate the campus to a low-risk area. Given the nature and severity of the potential hazards, mitigation options other than relocation were judged infeasible.

Once the decision was made to relocate, the district went to work to identify an alternate site for the school facilities. The selected location for the campus was completely out of the dam inundation area and far removed from the high-pressure oil pipelines. Thus the risk posed by the dam and oil pipelines hazards would be eliminated. Although the campus would still be within an active earthquake fault area, the new campus buildings would be constructed to fully conform to 1995 building code provisions, thus making them more resistant to seismic damage than the buildings being replaced.

The district then agreed to turn the land over to the Newhall County Water District as soon as the relocation effort was underway. The old school property is located above two active wells, which the water district can use to supply their customers in Castaic. In doing so, they changed the property deed to restrict human habitation and development and to return the site to natural open space.

The Castaic School District financed the relocation effort through a combination of grant money from FEMA and the sale of bonds. The district applied for and received a $7.2 million grant through FEMA's Hazard Mitigation Grant Program for the market value of the property, including the existing structures and infrastructure. The district used this funding, plus $20 million generated by school bonds, to rebuild the elementary school, district office, and middle school, and to relocate the elementary school students into temporary buildings during the construction of the new facilities. The new middle school opened in the fall of 1996 and the new elementary school opened in August 1997.

## CASE STUDY: VIRGIN ISLANDS BUILDING CODE

On September 18, 1989, Hurricane Hugo, a Category 4 storm, passed over the Virgin Islands with sustained winds of 130 mph, leaving near-total devastation in its wake. Losses of $1.5 billion included damage or destruction of 95 percent of the buildings and 90 percent of the power supply. Almost all public buildings, including hospitals, schools, and shelters, sustained major damage or were destroyed. The tourist industry was in a shambles. All communications with Puerto Rico and the mainland were severed. A Presidential disaster declaration was announced.

The Government of the Virgin Islands, with support from FEMA, began an immediate effort to identify measures to mitigate damage from future storms. Projects identified included upgrading the building codes and building practices, training building inspectors, initiating projects to harden the power grid, and establishing public education programs to show residents how to perform simple mitigation measures and their value.

With technical assistance from FEMA, a new building code was written and implemented. The code required anchoring systems, hurricane clips, shutters, and other measures to hold buildings together and reduce flying debris. Piers, water production, distribution, and oil storage facilities were strengthened. A massive public education program was launched.

When Hurricane Marilyn hit, the public buildings performed well, but most single-family homes lost their roofs. Once again, the building codes were amended to strengthen the quality of residential construction. The Governor's Office initiated a comprehensive program to repair damaged roofs. The Home Protection Roofing Program provided more than 350 homeowners with roofs to withstand a Category 2 storm.

Hurricane Georges, occurring in September 1998, packing winds of more than 100 mph, put these measures to the test. The results were excellent. Public and private efforts had retrofitted or rebuilt most of the structures on the island by September 1998. Damage to homes was limited to less than 2 percent of the islands. All hotels survived with little or no damage. Power was interrupted to 15 percent of the island but was fully restored within two weeks. Schools and other public structures were undamaged and provided safe havens for the residents to ride out the storm. Officials attribute the reduction in damages not just to the stronger code but also to the intensive education effort for building officials, contractors, and building owners about proper building practices and other mitigation strategies.

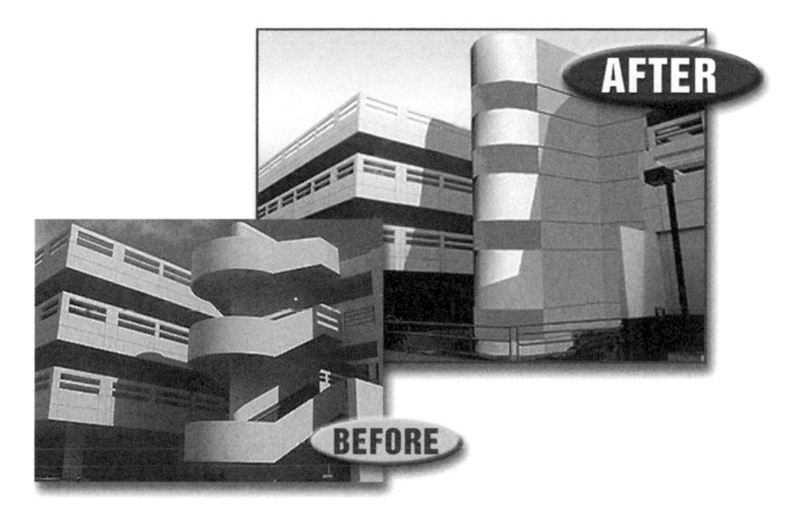


**Figure 3-3** Guam Memorial Hospital before and after rebuilding and adding mitigation techniques for high wind. FEMA Photo.

## CASE STUDY: ARNOLD, MISSOURI

The city of Arnold, Missouri, is located about 20 miles southwest of St. Louis at the confluence of the Meramec and Mississippi Rivers. The geography of Arnold causes it to be impacted from backwaters from the Mississippi and direct flooding of the Meramec and its tributaries. The floodplains of both these rivers had experienced extensive development. Because of these concerns, Arnold had adopted a floodplain management program in 1991; however, it had no storm water management program.

The Midwest floods of the spring and summer of 1993 resulted in record flood losses and damages totaling between $12 to $16 billion. Nine states, 532 counties, and more than 55,000 homes were flooded. The 1993 floods had a devastating effect on the 18,000 residents of Arnold. Approximately 250 structures were under water, and more than 528 households applied for disaster assistance, which amounted to more than $2 million. The city had to operate more than 60 sandbag sites to hold off the waters. Parts of the town were under water for up to two weeks.

When the water receded, the city of Arnold started an aggressive program to voluntarily buy out properties in the floodplain. It proposed the purchase of single-family homes, commercial structures, and mobile homes. It developed a plan to turn the purchased land into an open space greenway along the west banks of the Meramec and Mississippi Rivers. They initiated a public education campaign for the purchase of flood insurance because only 208 of the 908 floodplain properties had flood insurance.

Although they were unable to implement their 1991 flood plain management plan, their commitment to mitigation paid off. Arnold received significant HMGP funding for their buyout because of their commitments. By combining HMGP, community development block grant (CDBG), and other HUD funding, it proceeded with its buyout program. Initial estimates put the program costs at $3.5 million, but in the end it would cost $7.3 million.

**Figure 3-4** An example of relocation of homes out of the floodplain. FEMA Photo.

In the midst of this effort, Arnold experienced another major flood in 1995. The 1995 flood was the fourth largest flood in Arnold's history, but this time the results were dramatically different. Only four sandbag sites were needed, only 26 households applied for assistance, and the damage costs were less than $40,000.

Arnold continued its buyout program into 1996, working to obtain funding to remove the last 34 properties. The city continues to make other structural changes, including bridge elevations to restore the floodplain to its natural state and to provide a buffer for any future flooding.

# 4. The Disciplines of Emergency Management: Response

## INTRODUCTION

When a disaster event such as a flood, earthquake, or hurricane occurs, the first responders to this event are always local police, fire, and emergency medical personnel. Their job is to rescue and attend to those injured, suppress fires, secure and police the disaster area, and to begin the process of restoring order. They are supported in this effort by local emergency management personnel and community government officials.

If the size of the disaster event is so large that the capabilities of local responders are overwhelmed and the costs of the damage inflicted exceed the capacity of the local government, the mayor or county executive will turn to the governor and state government for assistance in responding to the event and in helping the community to recover. The governor will turn to the state's emergency management agency and possibly the State National Guard and other state resources to provide this assistance to the stricken community.

If the governor decides, based on information generated by community and state officials, that the size of the disaster event exceeds the state's capacity to respond, the governor will make a formal request to the President for a Presidential major disaster declaration. This request is prepared by state officials in cooperation with regional staff from FEMA. The governor's request is analyzed first by the FEMA Regional Office and then forwarded to FEMA headquarters in Washington, D.C. With a recommendation FEMA headquarters staff review and evaluate the governor's request and forward their analysis and recommendation to the President. The President considers FEMA's recommendation and then makes a decision to grant the declaration or to turn it down.

If the President grants a major disaster declaration, FEMA activates the Federal Response Plan (FRP) and proceeds to direct 27 federal departments and agencies including the American Red Cross in support of state and local efforts to respond to and recover from the disaster event. The Presidential declaration also makes available several disaster assistance programs through FEMA and other federal agencies designed to assist individuals and communities to begin the process of rebuilding their homes, their community infrastructure, and their lives.

When a major disaster strikes in the United States, the aforementioned chronology describes how the most sophisticated and advanced emergency management system in the world responds and begins the recovery process. This system is built on coordination and cooperation among a significant number of federal, state, and local government agencies, volunteer organizations, and, more recently, the business community.

In the 1990s the emergency management system in the United States was tested repeatedly by major disaster events such as the 1993 Midwest floods, the 1994 Northridge, California, earthquake, and a series of devastating hurricanes and tornadoes. In each instance, the system worked to bring the full resources of the federal, state, and local governments to produce the most comprehensive and effective response possible. The system also leveraged the capabilities and resources of America's cadre of volunteer organizations to provide immediate food and shelter. In recent years, government officials and agencies at all levels have begun to reach out to the business community to both leverage their response capabilities and to work closer with them in the recovery effort.

The September 11 terrorist attacks have caused all levels of government to reevaluate response procedures and protocols. The unusual loss of so many first responders to this disaster event has resulted in numerous after-action evaluations that will likely lead to changes in the procedures and protocols for first responders in the future. Additionally, the possibility of future terrorism attacks has focused attention on how best to protect first responders from harm in future attacks. These issues are discussed in detail in Chapter 9.

This chapter describes how local, state, and federal government officials and their partners respond to disasters in this country. The chapter includes sections discussing local response, state response, volunteer groups response, the Incident Command System, the FRP, and communications among responding agencies.

## LOCAL RESPONSE

Minor disasters occur daily in communities around the United States. Local fire, police, and emergency medical personnel respond to these events usually in a systematic and well-planned course of action. Firefighters, police officers, and emergency medical technicians respond to the scene. Their job is to secure the scene and maintain order, rescue and treat those injured, contain and suppress fire or hazardous conditions, and retrieve the dead.

The types of minor disasters responded to at the community level include hazardous materials transportation and storage incidents, fires, and localized flooding. Local officials are also the first responders to major disaster events such as large floods, hurricanes, and major earthquakes, but in these instances their efforts are supported, upon request by community leaders, by state government and, by request of the governor and approval of the President, by the federal government.

The actions of local first responders are driven by procedures and protocols developed by the responding agency (i.e., fire, police, and emergency medical). Most communities in the United States have developed communitywide emergency plans that incorporate these procedures and protocols. These community emergency plans also identify roles and responsibilities for all responding agencies and personnel for a wide range of disaster scenarios. These plans also include copies of the statutory authorities that provide the legal backing for emergency operations in the community.

In the aftermath of the September 11 terrorist events, many communities are reviewing and reworking their community emergency plans to include procedures and protocols for responding to all forms of terrorist attacks, including bioterrorism and weapons of mass destruction.

## First Responder Roles and Responsibilities

The roles and responsibilities of first responders are often detailed in the community emergency plan. A review of the Madison County, North Carolina, All-Hazard Plan provides a typical example of the contents of community emergency plans and the designation of roles and responsibilities among local first responders.

### CONTENTS OF MADISON COUNTY (NC) ALL-HAZARD PLAN

- Instruction for use
- Basic plan
- Glossary
- Acronyms and abbreviations
- Laws and ordinances
- Madison County Emergency Management Ordinance
- Madison County State of Emergency Ordinance
- Proclamation of State of Emergency
- Proclamation of Terminating
- Mutual aid
- Madison County Operation Plan (assignment of responsibilities)
    - Chairperson, County Commissioners
    - County Manager
    - Finance
    - Emergency Management Coordinator
    - Radiological Officer
    - Damage Assessment Officer (tax assessor)
    - Sheriff
    - Towns
    - County Fire Marshal and Fire Chiefs
    - Incident Commander
    - EMS Coordinator
    - Social Services Director
    - Amateur Radio Emergency Service
    - Health Director
    - Medical Center Disaster Coordinator
    - Medical Examiner

*continues*

- Mental Health Coordinator
- Superintendent of Schools
- American Red Cross
- Public Works
- Salvation Army
- Direction and control
- Communications
- Notification and warning
- Emergency public information
- Law enforcement
- Fire and rescue
- Public work/landfill
- Health and medical services
- Evacuation and transportation
- Shelter and mass care, including Red Cross
- Damage assessment/recovery
- Radiological protection
- Resource management
- Nuclear threat/hazard
- Hazardous Materials Southern Railway (ATT) EOC
- Hurricanes and flooding
- Transportation accidents
- Mass casualties
- Winter storms
- Tornadoes
- Civil disorders
- Dam failure
- Major incidents at public schools
- I-40 detour traffic
- Search and rescue plan
- 911 failure
- Power failure/countywide
- Formation of LEPC
- Contingency plan
- (ATT) EOC—Federal Response Plan—Southern Railroad/HAZ Plan

*Source:* Madison County All-Hazard Plan

## Local Emergency Managers

It is usually the responsibility of the designated local emergency manager to develop and maintain the community emergency plans. This individual often holds one or more other positions in local government such as fire or police chief and serves only part-time as the community's emergency manager. The profession of local emergency management has been maturing since the 1980s. There are now more opportunities for individuals to receive formal training in emergency management in the United States. Currently, more than 80 junior college, undergraduate, and graduate programs offer courses and degrees in emergency management and related fields. Additionally, FEMA's Emergency Management Institute (EMI) located in Emmitsburg, Maryland, offers emergency management courses on campus and through distance learning programs. EMI has also worked closely with junior colleges, colleges and universities, and graduate schools to develop coursework and curriculums in emergency management. More information on EMI and other emergency management education programs can be found in Chapter 6.

## THE CERTIFIED EMERGENCY MANAGER PROGRAM

The International Association of Emergency Managers (IAEM) created the Certified Emergency Manager© (CEM) Program to raise and maintain professional standards. It is an internationally recognized program that certifies achievements within the emergency management profession.

CEM certification is a peer review process administered through the International Association of Emergency Managers. You do not have to be an IAEM member to be certified, although IAEM membership does offer you a number of benefits that can assist you through the certification process. Certification is maintained in five-year cycles.

The CEM Program is served by a CEM Commission that is composed of emergency management professionals, including representatives from allied fields, education, the military, and private industry. Development of the CEM Program was supported by FEMA, the National Emergency Management Association (NEMA), and a host of allied organizations.

*Source:* IAEM, www.iaem.org

## ROLES AND RESPONSIBILITIES OF THE EMERGENCY MANAGEMENT COORDINATOR IN MADISON COUNTY ALL-HAZARD PLAN

Emergency Management Coordinator

a. Perform assigned duties according to state statutes and local ordinances.
b. Responsible for planning in accordance with federal and state guidelines and coordinating of emergency operations within the jurisdiction.

*continues*

c. Maintain current inventories of public information resources.
d. Ensure regular drills and exercises are conducted to test the functions of the EOP annually.
e. Identify resources county and private and maintain current inventories of county-owned resources, including sources and quantities, and develop mutual aid agreements to control these resources.
f. Request funding for maintaining equipment for radiation hazard evaluation and exposure control.
g. Establish and equip the County Emergency Operating Center (EOC) to include primary and backup radio communications (fixed and mobile), and provide for operations on a continuous basis as required.
h. Ensure adequate training for the emergency management organization.
i. Ensure means are available within the jurisdiction to gather necessary information (i.e., fuel storage facilities, major distributors, and end-user status), during the energy emergency status.
j. Provide emergency information materials for the public including non-English-speaking groups.
k. Prepare written statements of agreements with the media to provide for dissemination of essential emergency information and warning to the public, including the appropriate protective actions to be taken.
l. Coordinate exercises and tests of emergency systems within the jurisdiction.
m. Maintain liaison with utility companies to arrange for backup water, power, and telephone service during emergencies.
n. Maintain working relationships with the media and a current list of radio stations, television stations, and newspapers to be used for public information releases.
o. Alert and activate, as required by the County Emergency Management Organization, when informed of an emergency within the county.
p. Receive requests for assistance from municipalities within the county and direct aid to areas where needed.
q. Coordinate disaster assessment teams conducting field surveys.
r. Conduct a public information campaign to disseminate disaster assistance information as necessary.
s. Maintain listing of medical facilities.
t. Collect data and prepare damage assessment reports.
u. Provide for the storage, maintenance, and replenishment/replacement of essential equipment and materials (e.g., medical supplies, food and water, radiological instruments).
x. Develop a schedule for testing, maintaining, and repairing EOC and other emergency equipment.
y. Develop and maintain the EOC Standard Operating Guides, including an activation checklist and notification/recall roster.
z. Establish and maintain coordination with other jurisdictional EOCs as appropriate.

aa. Provide for adequate coordination of recovery activities among private, state, and federal agencies/organizations.
bb. Develop procedures to warn areas not covered by existing warning systems.
cc. Coordinate warning resources with neighboring counties.
dd. Develop and maintain a public information and education program.
ee. Assist the public information officer (PIO) in disseminating public information and education program.
ff. Identify and develop procedures for potential evacuation areas in accordance with the county's hazard analysis.
gg. Identify population groups requiring special assistance during evacuation (e.g., senior citizens, the very ill and disabled, nursing homes, prison population) and assure that they have evacuation procedures in place.
hh. Establish Disaster Assistance Centers if appropriate.
ii. Initiate the return of the population as soon as conditions are safe at the direction of the Chairman, Board of County Commissioners.
jj. Initiate the crisis upgrading and marking of shelters.
kk. Identify and survey congregate care shelter facilities that have lodging and mass feeding capabilities.
ll. Develop procedures to activate and deactivate shelters and ensure that ARC and DSS develop shelter SOGs.
mm. Establish public information and education programs on sheltering.
nn. Assist with designating facilities and arranging for the shelter needs of institutionalized or special needs groups.
oo. Designate shelter facilities in the reception area with the shortest commuting distance to the hazardous area for essential workers and their families.
pp. Appoint a Damage Assessment Officer to coordinate overall damage assessment operations.
qq. Recruit damage assessment team members.
rr. Secure resources to support and assist with damage assessment activities (e.g., maps, tax data, cameras, identification, report forms).
ss. Establish a Utilities Liaison to coordinate information flow between the EOC and affected utilities.
tt. Assist with identification and notification of applicants that may be eligible for Public Assistance programs.
uu. Develop a flood warning system for areas in the county subject to frequent flooding.
vv. Appoint a Radiological Officer or perform duties of that office.
ww. Acquire and provide radiological monitoring equipment.
xx. Coordinate overall radiological protection activities.
yy. Coordinate resource use under emergency conditions and provide a system to protect these resources.

*continues*

zz. Support the LEPC in maintaining liaison with facility emergency coordinators to ensure availability of current information concerning hazards and response to an incident.
aaa. Ensure a critique of incident responses to assess and update procedures as needed.
bbb. Serve as the Community Emergency Coordinator as identified in SARA, Title Jill.
ccc. Assist the area staff and the energy policy council in obtaining the essential data for implementation of contingency plans.
ddd. Assure coordination of planning efforts among jurisdictions (e.g., municipalities, counties, facilities), including the development of notification/warning, response, and remediation procedures for covered facilities.
eee. Ensure serviceability of radiological monitoring instruments.
fff. Alert all emergency support services to the dangers associated with technological hazards and fire during emergency operations.
ggg. Advise decision makers on the hazards associated with hazardous materials.

*Source:* Madison County All-Hazard Plan

More and more communities have designated emergency managers responsible for guiding response and recovery operations. Training and education programs in emergency management are expanding dramatically, resulting in a growing number of professionally trained and certified local emergency managers. The maturing of this profession can only lead to more effective and efficient local responses to future disaster events.

## STATE RESPONSE

Each of the 50 states and 6 territories that constitute the United States maintains a state government office of emergency management. The names of the office vary from state to state. For example, in California it is called the Office of Emergency Services (OES), in Tennessee it is the Tennessee Emergency Management Agency (TEMA), in North Carolina it is the Department of Emergency Management (DEM), and in Florida it is the Florida Division of Emergency Management. A full list of State Emergency Management Organizations is presented in Appendix C.

Also, where the emergency management office resides in state government varies from state to state. In California, OES is located in the Office of the Governor, in Tennessee, TEMA reports to the Adjunct General, and in Florida, the emergency management function is located in the Office of Community Affairs. National Guard Adjutant Generals manage state emergency management offices in more than half of the 56 states and territories. The remaining state emergency management offices are lead by civilian employees.

Funding for state emergency management offices comes principally from FEMA and state budgets. For years, FEMA has provided up to $175 million annually to states to fund state and local government emergency management activities. This money is used by state emergency management agencies to hire staff, conduct training and exercises, and purchase equipment. A segment of this funding is targeted for local emergency management operations as designated by the state. State budgets also provide funding for emergency management operations, but this funding historically has been inconsistent, especially in those states with minimal annual disaster activity.

The principal resource available to governors in responding to a disaster event in their state is the National Guard. The resources of the National Guard that can be used in disaster response include personnel, communications systems and equipment, air and road transport, heavy construction and earth-moving equipment, mass care and feeding equipment, and emergency supplies such as beds, blankets, and medical supplies.

Response capabilities and capacities are strongest in those states and territories that experience high levels of annual disaster activity. North Carolina is one of those states with high risk of hurricanes and floods. How the North Carolina Department of Emergency Management describes its response process on its Website provides an example of state response functions.

## RESPONSE BY THE NORTH CAROLINA DEPARTMENT OF EMERGENCY MANAGEMENT RESPONSE

The division's emergency response functions are coordinated in a proactive manner from the State Emergency Operations Center located in Raleigh. Proactive response strategies used by the division include:

- Area Commands that are strategically located in an impacted region to assist with local response efforts using state resources
- Central warehousing operations managed by the state that allow for immediate delivery of bottled water, ready-to-eat meals, blankets, tarps, and the like; field deployment teams manned by division and other state agency personnel that assist severely impacted counties coordinate and prioritize response activity
- Incident action planning that identifies response priorities and resource requirements 12 to 24 hours in advance

The State Emergency Response Team (SERT), which consists of top-level management representatives of each state agency involved in response activities, provides the technical expertise and coordinates the delivery of the emergency resources used to support local emergency operations.

*continues*

When resource needs are beyond the capabilities of state agencies, mutual aid from other unimpacted local governments and states may be secured using the Statewide Mutual Aid agreement or Emergency Management Assistance Compact. Federal assistance may also be requested through the Federal Emergency Response Team, which collocates with the SERT during major disasters.

*Source:* North Carolina Department of Emergency Management, www.dem.dcc.state.nc.us.

## VOLUNTEER GROUP RESPONSE

Volunteer groups are on the front line of any disaster response. National groups such as the American Red Cross and the Salvation Army roster and maintain local chapters of volunteers who are trained in emergency response. These organizations work with local, state, and federal authorities to address the immediate needs of disaster victims. These organizations provide shelter, food, and clothing to disaster victims who have lost their homes to disasters large and small.

In addition to the Red Cross and the Salvation Army, numerous volunteer groups across the country provide aid and comfort to disaster victims. The National Volunteer Organizations Against Disasters (NVOAD) consists of 34 national member organizations, 52 state and territorial VOADs, and a growing number of local VOADs involved in disaster response and recovery operations around the country and abroad. Formed in 1970, NVOAD helps member groups at a disaster location to coordinate and communicate in order to provide the most efficient and effective response. A list of the NVOAD member organizations is provided.

### LIST OF NVOAD MEMBER ORGANIZATIONS

Adventist Community Services, www.adventist.communityservices.org
American Radio Relay League, www.arrl.org
American Red Cross, www.redcross.org
American's Second Harvest, www.secondharvest.org
Ananda Marga Universal Relief Team, www.amurt.org
Catholic Charities USA, www.catholiccharitiesusa.org
Christian Disaster Response, www.cdresponse.org
Church of the Brethren, www.brethren.org
Church World Services, www.cwserp.org
Episcopal Relief and Development, www.er-d.org

Friends Disaster Service
Humane Society of the United States, www.hsus.org
International Relief Friendship Foundation, IRFFint@aol.com
International Aid, www.gospelcom.net/ia
Lutheran Disaster Response, www.elca.org/dcs/disaster
Mennonite Disaster Services, www.mds.mennonite.net
National Emergency Response Team, www.nert-usa.org
National Organization for Victim Assistance, www.try-nova.org
Nazarene Disaster Response, www.nazarenedisasterresponse.org
Northwest Medical Teams International, www.nwmti.org
The Phoenix Society for Burn Survivors, www.phoenix-society.org
The Points of Light Foundation, www.pointsoflight.org
Presbyterian Disaster Assistance, www.pcusa.org/pcusa/wmd/pda/index.html
REACT International, www.reactintl.org
The Salvation Army, www.salvationarmyusa.org
Society of St. Vincent de Paul, http://home.aol.com/svdpus
Southern Baptist Disaster Relief, www.namb.net/dr/pages/beginnings.asp
United Jewish Communities, www.ujcna.org
United Methodist Committee on Relief, http://gbgm-umc.org/umcor/emergency.stm
United States Service Command, www.ussc-hq.org
Volunteers of America, www.voa.org
World Vision, www.worldvision.org

*Source:* National Volunteer Organizations Against Disasters (NVOAD)

## GERMAN SALVATIONISTS PROVIDE AID TO FLOOD VICTIMS

August 15, 2002. Salvationists in Dresden, Germany, have been working tirelessly to help people affected by the flooding that has brought chaos to the city and much of the surrounding area. The River Elbe is already at its highest point since the mid-19th century and water levels are still rising. More than 3,000 people have so far been forced to evacuate their homes.

When the flooding started, Salvationists from the Salvation Army centre in Dresden immediately offered support to the emergency workers, as no official supplies were then being provided to fire and ambulance personnel. However, the main focus of attention quickly shifted to the victims of the flooding.

The Salvation Army corps (church) building in Dresden is located on high ground and, unlike many buildings in the city, still has power, so cooking and

continues

food preparation are possible. More than 2,000 meals have been provided so far. Local hotels and a bakery are assisting with preparation and two Salvation Army mobile kitchens are being used to deliver food.

There is great concern for the many elderly people who are unable to leave their properties but who now have no power for cooking or heating. In addition to providing hot soup, consideration is being given to assisting these elderly people to find alternative, temporary accommodation. Many offers of help have come in, including from a nurse who put herself forward to provide assistance after the hospital she was working in was evacuated.

Donations of clothing and supplies have, up to now, had to be turned down because of a lack of storage space. The Salvation Army's International Emergency Services Office has arranged for US$25,000 to be sent out from International Headquarters and these funds will be used, among other things, to hire a suitable warehouse where donations can be stored.

*Source:* Salvation Army

## INCIDENT COMMAND SYSTEM

A difficult issue in any response operation is determining who is in charge of the overall response effort. The Incident Command System (ICS) was developed after the 1970 fires in southern California. Duplication of efforts, lack of coordination, and communication hindered all agencies responding to the expanding fires. The main function of ICS is to establish a set of planning and management systems that would help the agencies responding to a disaster to work together in a coordinated and systematic approach. The step-by-step process enables the numerous responding agencies to effectively use resources and personnel to respond to those in need.

There are multiple functions in the ICS system. They include common use of terminology, integrated communications, a unified command structure, resource management, and action planning. A planned set of directives includes assigning one coordinator to manage the infrastructure of the response, assigning personnel, deploying equipment, obtaining resources, and working with the numerous agencies that respond to the disaster scene. In most instances the local fire chief or fire commissioner is the Incident Commander.

For the ICS to be effective, it must provide for effective operations at three levels of incident character: (1) single jurisdiction and/or single agency, (2) single jurisdiction with multiple agency support, and (3) multijurisdictional and/or multiagency support. The organizational structure must be adaptable to a wide variety of emergencies (i.e., fire, flood, earthquake, and rescue). The ICS includes agency autonomy, management by objectives, unity integrity, functional clarity, and effective span of control. The logistics, coordination, and ability of the multiple agencies to work together must adhere to the ICS so that efficient leadership is maintained during the disaster. One of the most significant problems before the ICS was that agencies who would respond to major disasters would assign their own commander and there would be power struggles, miscommunication, and duplication of efforts (Irwin, 1980).

There are five major management systems within the ICS. They include Command, Operations, Planning, Logistics, and Finance.

- The *Command Section* includes developing, directing, and maintaining communication and collaboration with the multiple agencies on site, working with the local officials, the public, and the media to provide up-to-date information regarding the disaster.
- The *Operations Section* handles the tactical operations, coordinates the command objectives, and organizes and directs all resources to the disaster site.
- The *Planning Section* provides the necessary information to the command center to develop the action plan to accomplish the objectives. This section also collects and evaluates information as it is made available.
- The *Logistics Section* provides personnel, equipment, and support for the Command Center. They handle the coordination of all services that are involved in the response, from locating rescue equipment to coordinating the response for volunteer organizations such as the Salvation Army and the Red Cross.
- The *Finance Section* is responsible for accounting for funds used during the response and recovery aspect of the disaster. The finance section monitors costs related to the incident and provides accounting procurement time recording cost analyses.

In today's world, the public, private, and political values at risk in major emergencies demand the most efficient methods of response and management. Meeting this demand when multiple and diverse agencies are involved becomes a difficult task. The Unified Command concept of ICS offers a process that all participating agencies can use to improve overall management, whether their jurisdiction is of a geographical or functional nature (Irwin, 1980).

The Unified Command is best used when there is a multiagency response. Because of the nature of the disaster, multiple government agencies need to work together to monitor the response and manage the large number of personnel who respond to the scene (see Figure 4-1). It allows for the integration of the agencies to operate under one overall response management.

## PROCEDURES FOR INCIDENT COMMAND SYSTEM

For an ICC to be effective, procedures need to be followed closely:

- A command post needs to be established.
- Proper equipment, such as computers, radios, and telephone lines, needs to be installed and in working order.

*continues*

- A media/press area needs to be established.
- Topographic maps need to be located and posted. After tornadoes, street signs or other identifying landmarks are destroyed and rescue personnel are unable to use traditional road maps.
- Locate/prepare a missing persons lists.
- Monitor the movement and location of triage areas and transportation of victims.
- Have the ability to maintain continuous communication with local hospitals to monitor the number of victims received.
- Establish and grid the search area.
- Based on the type of disaster, such as flooding, responders may have to use boats to search for and rescue victims.
- Determine what resources are available within the local area and what resources are being deployed.
- As the response system expands, reevaluate tasks that need to be performed and develop new tasks.

The Incident Commander (IC) prepares to delegate responsibilities as needed, to maintain focus on the overall situation. The IC needs to assign positions, such as debriefers, coordinators, and unit leaders, to manage the command center. As the response and recovery process proceeds, the IC needs to have an ongoing dialogue with staff and officials to monitor and manage the response. The IC needs to evaluate the continuing needs of the responders and determine if additional resources are needed. In the after-action reports, discussion and evaluation of the disaster determines the success based on the initial competence and effectiveness of the Incident Commander and the Center.

## THE FEDERAL RESPONSE

Once the governor has determined that a disaster event has overwhelmed the capacity of state and local governments to effectively respond and to subsequently fund the recovery effort, the governor forwards a letter to the President requesting a Presidential disaster declaration. This is the first step toward involving federal officials, agencies and departments, and resources in a disaster event. If the event is declared a major disaster by the President, 27 federal departments and agencies, including the American Red Cross, work together to support the efforts of state and local officials.

FEMA is responsible for coordinating all federal activities in support of state and local response and recovery efforts in a presidentially declared disaster. In such an instance, FEMA activates the Federal Response Plan (FRP). FEMA also manages several programs that provide disaster assistance to individuals and affected communities. These programs are discussed in detail in Chapter 5.

**Figure 4-1** Wilmington, NC, September 16, 1999 — Crews from all local county and cities man the (New Hanover) County Emergency Operations Center in Wilmington, NC coordinating all disaster information related to the effects of and responses to Hurricane Floyd. Photo by DAVE GATLEY/ FEMA News Photo

## Presidential Disaster Declaration Process

The Presidential disaster declaration makes available the resources of the federal government to the disaster area. Although a formal declaration does not have to be signed for the federal government to respond, the governor must make a formal request for assistance and specify in the request the specific needs of the disaster area. The Presidential major disaster declaration process is provided as follows.

## PRESIDENTIAL MAJOR DISASTER DECLARATION PROCESS

A disaster declaration should include the following guidelines:

- Contact is made between the affected state and the FEMA regional office. This contact may take place before or immediately following the disaster.
- If it appears the situation is beyond state and local capacity, the state requests FEMA to conduct a joint Preliminary Damage Assessment (PDA). Participants in the PDA will include FEMA, state, and local government representatives and other federal agencies.
- Based on the PDA findings, the governor submits a request to the President through the FEMA Regional Director for either a major disaster or an emergency declaration and identifying the counties impacted.
- The FEMA Regional Office submits a summary of the event and a recommendation based on the results of the PDA to FEMA headquarters, along with the governor's request.
- Upon receipt of these documents, headquarters senior staff convenes to discuss the request and determine the recommendation to be made to the President.
- FEMA's recommendation is forwarded to the White House for review.
- The President declares a major disaster or an emergency.

*Source:* Federal Response Plan, April 1999

The decision to make a disaster declaration is completely at the discretion of the President. There are no set criteria to follow and no government regulations to guide which events are declared by the President and which events are not. FEMA has developed several factors it considers in making its recommendation to the President, including individual property losses per capita, level of damage to existing community infrastructure, and insurance coverage. In the end, however, the decision to make the declaration is the President's alone.

A Presidential disaster declaration can be made in as short a time as a few hours, as was the case in the 1994 Northridge earthquake and the 1995 Oklahoma City bombing. Sometimes it takes weeks for damages to be assessed and the capability of state and local jurisdictions to fund response and recovery efforts to be evaluated. If the governor's request is turned down by the President, the governor has a right to appeal and can be successful, especially if new damage data become available and are included in the appeal.

Presidential declarations are routinely sought for such events as large floods, hurricanes, earthquakes, and big tornadoes. In recent years, governors have become more inventive and have requested Presidential disaster declarations for snow removal, drought, West Nile Virus, and economic losses caused by failing industries such as the Northwest salmon spawning decline.

Since 1976, there have been 906 Presidential disaster declarations, averaging 34 declarations per year (see Table 4-1). As an example of disaster declaration activity in a single year, in 1999 there were 50 major disaster declarations in 38 states, including 18 hurricanes (13 alone for Hurricane Floyd), 11 tornadoes, 7 floods, 6 winter storms, 6 severe storms, 1 flash flood, and 1 winter freeze (see Table 4-2).

**Table 4-1** Total Major Disaster Declarations, 1976–2001

| Year | Total Disaster Declarations |
|---|---|
| 1976 | 30 |
| 1977 | 22 |
| 1978 | 25 |
| 1979 | 42 |
| 1980 | 23 |
| 1981 | 15 |
| 1982 | 24 |
| 1983 | 21 |
| 1984 | 34 |
| 1985 | 27 |
| 1986 | 28 |
| 1987 | 23 |
| 1988 | 11 |
| 1989 | 31 |
| 1990 | 38 |
| 1991 | 43 |
| 1992 | 45 |
| 1993 | 32 |
| 1994 | 36 |
| 1995 | 32 |
| 1996 | 75 |
| 1997 | 44 |
| 1998 | 65 |
| 1999 | 50 |
| 2000 | 45 |
| 2001 | 45 |
| **Total** | **906** |
| **Average** | **34** |

*Source:* FEMA, www.fema.gov

## Federal Response Plan

In 1992, FEMA developed the FRP. FEMA defines the FRP as a

> Signed agreement among 27 Federal departments and agencies, including the American Red Cross, that: Provides the mechanism for coordinating delivery of Federal assistance and resources to augment efforts of State and local governments overwhelmed by a major disaster or emergency, Supports implementation of the Robert T. Stafford Disaster Relief and Emergency Assistance Act, as amended (42 U.S.C. 5121, et seq.), as well as individual agency statutory authorities and Supplements other Federal emergency operations plans developed to address specific hazards.

**Table 4-2** FEMA Major Disaster Activity, January 1, 1999, to December 31, 1999

| Date | State | Incident |
|---|---|---|
| 01/15 | Tennessee | Winter Storm |
| 01/15 | Alabama | Winter Storm |
| 01/19 | Tennessee | Tornado |
| 01/21 | Maine | Winter Storm |
| 01/21 | Louisiana | Winter Storm |
| 01/23 | Arkansas | Tornadoes |
| 01/25 | Mississippi | Winter Storm |
| 02/09 | California | Winter Freeze |
| 02/17 | Wyoming | Winter Storm |
| 04/09 | Louisiana | Tornadoes |
| 04/20 | Missouri | Floods |
| 04/20 | Georgia | Tornadoes |
| 05/04 | Oklahoma | Tornadoes |
| 05/04 | Kansas | Tornadoes |
| 05/06 | Texas | Tornadoes |
| 05/12 | Tennessee | Tornadoes |
| 05/17 | Colorado | Severe Storms |
| 05/21 | Iowa | Floods |
| 05/28 | Illinois | Severe Storms |
| 06/08 | North Dakota | Severe Storms |
| 06/09 | South Dakota | Severe Storms |
| 07/20 | Nevada | Flash Floods |
| 07/22 | Iowa | Floods |
| 07/28 | Minnesota | Floods |
| 08/16 | Wisconsin | Severe Storms |
| 08/16 | Utah | Tornado |
| 08/20 | Nebraska | Floods |
| 08/22 | Texas | Hurricane Bret |
| 08/26 | Minnesota | Severe Storms |
| 09/01 | Pennsylvania | Floods |
| 09/07 | Virginia | Tornado |
| 09/09 | North Carolina | Hurricane Dennis |
| 09/16 | North Carolina | Hurricane Floyd |
| 09/18 | Virginia | Hurricane Floyd |
| 09/18 | Pennsylvania | Hurricane Floyd |
| 09/18 | New Jersey | Hurricane Floyd |
| 09/19 | New York | Hurricane Floyd |
| 09/21 | Delaware | Hurricane Floyd |
| 09/22 | Pennsylvania | Hurricane Dennis |
| 09/21 | South Carolina | Hurricane Floyd |
| 09/22 | Florida | Hurricane Floyd |
| 09/22 | New Mexico | Floods |
| 09/23 | Connecticut | Hurricane Floyd |
| 09/24 | Maryland | Hurricane Floyd |
| 10/15 | Arizona | Severe Storms |
| 10/18 | New Hampshire | Hurricane Floyd |
| 10/20 | Florida | Hurricane Irene |
| 11/11 | Vermont | Hurricane Floyd |
| 11/18 | Maine | Hurricane Floyd |
| 11/23 | U.S. Virgin Islands | Hurricane Lenny |

*Source:* FEMA, www.fema.gov

The fundamental goal of the FRP is to maximize available federal resources in support of response and recovery actions taken by state and local emergency officials.

## TYPES OF FEDERAL ASSISTANCE AVAILABLE

The FRP makes available the following types of assistance:
*To deliver immediate relief:*

- Initial response resources, including food, water, emergency generators
- Emergency services to clear debris, open critical transportation routes, provide mass sheltering and feeding

*To speed return to normal and reduce damage from future occurrences:*

- Loans and grants to repair or replace damaged housing and personal property
- Grants to repair or replace roads and public buildings, incorporating to the extent practical hazard-reduction structural and nonstructural measures
- Technical assistance to identify and implement mitigation opportunities to reduce future losses
- Other assistance, including crisis counseling, tax relief, legal services, job placement

*Source:* Federal Response Plan

Four operating principles are integral in the successful implementation of the FRP:

1. As stated in FEMA's definition of the FRP, the FRP is "the mechanism for coordinating delivery of Federal assistance to augment efforts of State and local governments overwhelmed by a major disaster or emergency." The key word is *augment*. The FRP does not call for FEMA or any other federal agency to take over direction and control of a disaster relief effort; no governor would allow that to happen. On the contrary, and in contrast to public perception on occasion, the FRP defines how FEMA and its fellow federal agencies and the American Red Cross will support state and local governments in the disaster relief effort. Direction and control of all presidentially declared major disasters remains in the hands of the governor and local officials.
2. The FRP is the only working agreement in existence that involves one federal agency directing the activities of a large number of other federal agencies. This agreement is truly unique in U.S. government. Basically all signatories to the FRP agree to follow the direction of FEMA in providing disaster

assistance from their respective agencies. Nowhere else in the federal government are the full resources of 26 federal agencies brought to bear on a single civil problem.

3. The FRP includes a series of agreements between FEMA and the participating federal agencies and the American Red Cross that clearly define the types of services and resources that FEMA expects each agency to be able to provide in the event of a Presidential declaration. In other words, each participating agency agrees to retain the capability to deploy personnel, services, and other resources on a 24-hour on-call basis. Also, agencies are expected to have contract vehicles in place that allow for rapid procurement of contractor services and products. This ensures that when FEMA requests services and resources from a participating agency, they are available and ready to go immediately.
4. What really makes the FRP work is money. FEMA pays for everything out of its annual Disaster Relief Fund (DRF) or from supplemental funding made available for major catastrophes by Congress. FEMA receives approximately $325 million annually for the DRF from Congress and uses these funds to pay for services and products provided by FRP partner agencies. This means that FRP partner agencies do not have to tap existing budgets that are already programmed for disaster spending authority and ensures that these agencies will respond quickly to the direction of FEMA to support state and local efforts.

The way it works is that FEMA, through the information and planning function, identifies state and local needs and matches these needs to one or more of the FRP partner agencies to address these needs. FEMA "mission assigns" specific tasks to those agencies capable of completing these tasks, and it is already understood that FEMA will reimburse the FRP partner agencies per a preexisting agreement concerning costs of services.

For example, debris removal is a critical first step in helping a community on the road to recovery. The state will request assistance from FEMA, and FEMA will mission assign the U.S. Army Corps of Engineers to contract for debris removal services. The Corps executes a fast-track procurement with previously certified debris removal contractors, and work proceeds within days of the disaster event. The Corps will then bill FEMA for reimbursement for the costs of contracting with the debris removal contractor.

Another example would be FEMA mission assigning the Department of Defense (DOD) to provide helicopter support for rescue operations in a disaster area. By agreement, DOD helicopters are available for service, and the costs of flying and maintaining the helicopters and the salaries of the pilots and mechanics involved in the mission assignment will be reimbursed to DOD by FEMA. The DOD is a critical partner in the FRP. In fact, the DOD is the only FRP partner agency that maintains an office in FEMA headquarters staffed by a colonel who serves as a liaison between the DOD and FEMA. The full list of FRP partner agencies is presented as follows.

## FEDERAL RESPONSE PLAN PARTNERS

Department of Agriculture
Department of Commerce
Department of Defense
Department of Education
Department of Energy
Department of Health and Human Services
Department of Housing and Urban Development
Department of the Interior
Department of Justice
Department of Labor
Department of State
Department of Transportation
Department of the Treasury
Department of Veterans Affairs
Agency for International Development
American Red Cross
Environmental Protection Agency
Federal Communications Commission
Federal Emergency Management Agency
General Services Administration
National Aeronautics and Space Administration
National Communications System
Nuclear Regulatory Commission
Office of Personnel Management
Small Business Administration
Tennessee Valley Authority
U.S. Postal Service

*Source:* Federal Response Plan

### ESF Primary Agencies

Each of these FRP partners serves as a primary agency or support agency in one or more of the 12 Emergency Support Functions (ESFs) in the FRP. FEMA defines primary and support agencies as follows:

1. Orchestrating the federal agency support within the functional area for an affected state
2. Providing an appropriate level of staffing for operations at FEMA headquarters, the Regional Operations Center (ROC), Disaster Field Office (DFO), and Disaster Recovery Center (DRC)
3. Activating and subtasking support agencies

4. Managing mission assignments and coordinating tasks with support agencies, as well as appropriate state agencies
5. Supporting and keeping other ESFs and organizational elements informed of ESF operational priorities and activities
6. Executing contracts and procuring goods and services as needed
7. Ensuring financial and property accountability for ESF activities
8. Supporting planning for short- and long-term disaster operations

### ESF Support Agencies

When an ESF is activated in response to a disaster, each support agency for the ESF has operational responsibility for the following:

9. Supporting the ESF primary agency when requested by conducting operations using its authorities, cognizant expertise, capabilities, or resources
10. Supporting the primary agency mission assignments
11. Providing status and resource information to the primary agency
12. Following established financial and property accountability procedures

The 12 ESFs, a brief description of the activities conducted and managed in each ESF, and the identity of the primary agency in each ESF are provided as follows.

## EMERGENCY SUPPORT FUNCTIONS

The FRP employs a functional approach that groups under 12 ESFs the types of direct federal assistance that a state is most likely to need. Each ESF is headed by a primary agency designated on the basis of its authorities, resources, and capability in that functional area. Federal response assistance is provided using some or all ESFs as necessary. Federal ESF representatives coordinate with their counterpart state agencies.

**ESF #1: Transportation**, Department of Transportation. Assists federal agencies, state and local government entities, and voluntary organizations requiring transportation capacity to perform response missions.
**ESF #2**: **Communications**, National Communications System. Ensures the provision of federal telecommunications support to federal, state, and local response efforts.
**ESF #3: Public Works and Engineering**, U.S. Army Corps of Engineers, Department of Defense. Provides technical advice and evaluation; engineering services; contracting for construction management, inspection, and emergency repair of water and waste-water treatment facilities; and potable water and ice, emergency power, and real estate support to assist state(s) in lifesaving and life-protecting needs, damage mitigation, and recovery activities.
**ESF #4: Firefighting**, Forest Service, Department of Agriculture. Detects and suppresses wild land, rural, and urban fires resulting from, or occurring coincidentally with, a major disaster or emergency.

**ESF #5: Information and Planning**, Federal Emergency Management Agency. Collects, analyzes, processes, and disseminates information about a potential or actual disaster or emergency to facilitate the activities of the federal government in providing assistance to affected states.
**ESF #6: Mass Care**, American Red Cross. Coordinates federal assistance in support of state and local efforts to meet the mass care needs of victims, including sheltering, feeding, emergency first aid, and bulk distribution of emergency relief supplies.
**ESF #7: Resource Support**, General Services Administration. Coordinates provision of equipment, materials, supplies, and personnel to support disaster operations.
**ESF #8: Health and Medical Services**, Department of Health and Human Services. Provides coordinated federal assistance to supplement state and local resources in response to public health and medical care needs.
**ESF #9: Urban Search and Rescue**, Federal Emergency Management Agency. Deploys components of the National Search and Rescue Response System to provide specialized lifesaving assistance to state and local authorities, including locating, extricating, and providing initial medical treatment to victims trapped in collapsed structures.
**ESF #10: Hazardous Materials**, Environmental Protection Agency. Provides federal support to state and local governments in response to an actual or potential discharge/release of hazardous substances.
**ESF #11: Food**, Food and Nutrition Service, Department of Agriculture. Identifies, secures, and arranges for transportation of food assistance to affected areas requiring federal response, and authorizes the issuance of disaster food stamps.
**ESF #12: Energy**, Department of Energy. Helps restore the nation's energy systems following a major disaster requiring federal assistance; and coordinates with federal and state officials to establish priorities for repair of energy systems and to provide emergency fuel and power.

*Source:* FEMA, www.fema.gov

As noted earlier, the federal response is triggered by a Presidential disaster declaration. The response process and the structures activated by the FRP are provided as follows.

## THE FEDERAL RESPONSE PROCESS AND STRUCTURES OF THE FRP

When a disaster strikes, local and state responders work closely with volunteer agencies to respond to a major disaster. The mayor or county executive activates the local Emergency Operations Center (EOC). Upon request from local

continues

executives, the governor activates the state EOC, declares a state emergency or disaster, activates the state emergency operations plan, and begins an assessment.

If, from early damage reports, the state concludes that effective response may exceed the community's and state's resources, the state can request that FEMA regional officials join in conducting Preliminary Damage Assessments (PDAs). Data gathered in these assessments are used in the Presidential declaration process to determine the impact and magnitude of the damage and types of assistance needed, and to document that the disaster is beyond local and state capabilities to respond.

After the PDA teams have finished their work, the governor will determine whether to request federal disaster assistance. In order to make the request, the state must include specific information required by law and must guarantee that the cost-sharing provisions will be met. The governor's request for federal disaster assistance is addressed to the President and forwarded to the appropriate FEMA Regional Director, who evaluates the request and forwards a recommendation to the FEMA Director. The FEMA Director's recommendation is then forwarded to the President.

When the President determines that a state requires federal assistance, a formal disaster declaration is made, federal agencies utilize the FRP to meet the state's requests for assistance, and the federal disaster assistance process begins. The FRP enables a flexible and rapid federal response because the ESFs are selectively used.

## Structures of the FRP

The FEMA Operations Center (FOC) serves as FEMA's official notification point. This facility maintains a 24-hour capability to monitor all sources of information. Each region is supported by a Mobile Emergency Response Support (MERS) detachment and a Mobile Operations Center (MOC) that operates on a 24-hour basis to share information with the FOC.

In some instances, federal personnel representing some or all of the ESFs may be activated even before a disaster occurs. The Emergency Response Team Advance Element (ERT-A) is a federal group composed of FEMA staff and ESF representatives that respond initially in the field to assess the situation and begin response activities.

When a major declaration occurs (or before, for predictable events), the Regional Operations Center (ROC) is activated by the regional director at a FEMA regional office. The ROC is the initial coordination point for federal response efforts. If the President declares a disaster, a Disaster Field Office (DFO) will be set up near the disaster site to support response and recovery operations. A Federal Coordinating Officer (FCO) is appointed by the President (authority delegated to the Director, FEMA) and assumes coordination responsibilities. These include overseeing establishment of field offices, as necessary;

assessing types of relief most urgently needed; interacting with the State Coordinating Officer (SCO) to respond to state needs; and coordinating federal response and recovery activities. A federal interagency Emergency Response Team (ERT) is activated to support the FCO in coordinating overall federal disaster operation. The SCO is an individual appointed by the governor to coordinate state response and recovery operations with the federal government. This requires identifying requirements, including unmet needs and evolving support requirements, as well as coordinating other response and recovery activities with the state.

If a disaster is catastrophic and demands the full capability of FEMA, such as Hurricane Andrew, FEMA will deploy a team from FEMA Headquarters called the National Emergency Response Team (ERT-N) to coordinate FEMA resources.

Other elements of a federal response may include the Emergency Support Team (EST) and the Catastrophic Disaster Response Group (CDRG). The EST is an interagency group that provides coordination and operations support activities from the FEMA National Interagency Emergency Operations Center (NIEOC) at FEMA Headquarters. The CDRG is a group of representatives from all FRP signatory departments and agencies that operate as the national-level policy support forum.

*Source:* FEMA, www.fema.gov

## Urban Search and Rescue

As defined by FEMA, "urban search-and-rescue (US&R) involves the location, rescue (extrication), and initial medical stabilization of victims trapped in confined spaces. Structural collapse is most often the cause of victims being trapped, but victims may also be trapped in transportation accidents, mines and collapsed trenches." Urban search-and-rescue is considered a multihazard discipline because it may be needed for a variety of emergencies or disasters, including earthquakes, hurricanes, typhoons, storms and tornadoes, floods, dam failures, technological accidents, terrorist activities, and hazardous materials releases. The events may be slow in developing, as in the case of hurricanes, or sudden, as in the case of earthquakes (FEMA, www.fema.gov).

In the 1990s, FEMA established 28 US&R Task Forces all over the country. FEMA has set the policy for the establishment of these task forces and provides funding for equipping and training task force members. The US&R task forces are housed in local jurisdictions, and US&R team members are recruited from the police, fire and rescue, and emergency medical agencies within these jurisdictions. These 28 task forces can function anywhere in the United States. The task forces are on call 24 hours a day and can be ready for transport to a disaster site within 4 hours of a call-up.

## FEMA's URBAN SEARCH-AND-RESCUE TEAMS

**Q. What is FEMA's National US&R response system?**

A. This system is a framework for structuring local emergency personnel into integrated disaster response task forces. These task forces, complete with necessary tools and equipment, and specialized training and skills, are deployed by FEMA in times of catastrophic structural collapse.

**Q. How are FEMA US&R teams different from other search-and-rescue teams?**

A. FEMA teams organize existing search-and-rescue capability into a national program that can quickly deploy to an event. They have additional training, and must be able to deploy within 6 hours and to sustain themselves for 72 hours. They must also have a roster that fills 31 different positions with at least two people for each position. To receive the FEMA certification, the team must be approved by a US&R oversight board that includes leaders in the field and FEMA officials. One of the difficulties in obtaining the certification is being able to staff a complete roster of at least 62 trained individuals.

**Q. What kinds of positions make up the 31 in each team?**

A. First, all team members are trained and certified emergency medical technicians. Then positions fall into roughly four categories: search and rescue, medical, technical, and logistics. The search-and-rescue positions include engineers with expertise in shoring up, bracing, evaluating, breaching, and lifting structural components, rescue specialists, and search specialists who use trained and credentialed search dogs, cameras, and listening devices. The medical positions include physicians, EMTs, nurses, and others who can set up and staff a mobile field hospital. Technical positions include hazard materials specialists and communications specialists, among others.

**Q. What are the first steps the teams take when they arrive at a site?**

A. The FEMA US&R team meets with the field incident commander—the local firefighter or emergency specialist who is in charge of the site. After a general situation update and briefing, some team members set up a base of operations at the site, including tents, equipment, and a stage area. Meanwhile, search-and-rescue specialists and structural engineers inspect the site. They look for major problem areas, likely areas to search, the condition of the collapse, and hazardous materials. Also at this time, logistics team members are contacting local vendors to obtain heavy equipment, shoring materials, food, portable toilets, and other supplies.

**Q. Then what happens?**

A. The search-and-rescue specialists begin to gently and carefully move into the structure into areas that are not in imminent danger of collapse to get a better idea of the damage. They will have looked at blueprints of the building to understand its layout and will mark areas that need bracing and areas where victims can be seen. During this preliminary search, if any victim is found alive, the survey halts and stabilization efforts are concentrated there to get the victim out. After this preliminary search, the detailed search begins with dogs, cameras, and listening devices. Medical services are given to any victims who are found alive, so they are treated while they are being extricated.

Profile of a Rescue

While every search-and-rescue assignment is unique, a rescue might go something like this:

- Response always begins at the local level. Local rescuers always respond first. If the emergency is great enough, the state can request support from the FEMA task force.
- Following the disaster, the local emergency manager requests assistance from the state, the state in turn requests federal assistance, and FEMA deploys the three closest task forces.
- After arriving at the site, structural specialists, who are licensed professional engineers, provide direct input to FEMA task force members about the structural integrity of the building and the risk of secondary collapses.
- The search team ventures around and into the collapsed structure, shoring up structures and attempting to locate trapped victims. The team uses electronic listening devices, search cameras, and specially trained search dogs to locate victims.
- Once a victim is located, the search group begins the daunting task of breaking and cutting through thousands of pounds of concrete, metal, and wood to reach the victims. They also stabilize and support the entry and work areas with wood shoring to prevent further collapse.
- Medical teams, composed of trauma physicians, emergency room nurses, and paramedics, provide medical care for the victims as well as the rescuers, if necessary. A fully stocked mobile emergency room is part of the task force equipment cache. Medics may be required to enter the dangerous interior of the collapsed structure to render immediate aid.
- Throughout the effort, hazardous materials specialists evaluate the disaster site and decontaminate rescue and medical members who may be exposed to hazardous chemicals or decaying bodies.
- Heavy rigging specialists direct the use of heavy machinery, such as cranes and bulldozers. These specialists understand the special dangers of working in a collapsed structure and help ensure the safety of the victims and rescuers inside.
- Technical information and communication specialists ensure that all team members can communicate with each other and the task force leaders, facilitating search efforts and coordinating evacuation in the event of a secondary collapse.

*continues*

- Logistics specialists handle the more than 16,000 pieces of equipment to support the search and extrication of the victims. The equipment cache includes such essentials as concrete cutting saws, search cameras, medical supplies, and tents, cots, food, and water to keep the task force self-sufficient for up to four days.

*Source:* FEMA, www.fema.gov

Two of the US&R task forces have agreements with the U.S. Agency for International Development (USAID) to provide search-and-rescue services oversees. These two task forces are Metro-Dade Fire Department in Florida and the Fairfax County Fire & Rescue in Virginia. A full list of US&R task forces is presented in Table 4-3.

**Table 4-3** FEMA Urban Search-and-Rescue Task Forces

| State | Number | Organization |
|---|---|---|
| Arizona | AZ-TF1 | Phoenix, Arizona |
| California | CA-TF1 | LA City Fire Dept. |
| | CA-TF2 | LA County Fire Dept. |
| | CA-TF3 | Menlo Park Fire Dept. |
| | CA-TF4 | Oakland Fire Dept. |
| | CA-TF5 | Orange Co. Fire Authority |
| | CA-TF6 | Riverside Fire Dept. |
| | CA-TF7 | Sacramento Fire Dept. |
| | CA-TF8 | San Diego Fire Dept. |
| Colorado | CO-TF1 | State of Colorado |
| Florida | FL-TF1 | Metro-Dade Fire Dept. |
| | FL-TF2 | Miami Fire Dept. |
| Indiana | IN-TF1 | Marion County |
| Maryland | MD-TF1 | Montgomery Fire Rescue |
| Massachusetts | MA-TF1 | City of Beverly |
| Missouri | MO-TF1 | Boone County Fire Protection District |
| Nebraska | NE-TF1 | Lincoln Fire Dept. |
| Nevada | NV-TF1 | Clark County Fire Dept. |
| New Mexico | NM-TF1 | State of New Mexico |
| New York | NY-TF1 | NYC Fire and EMS, Police |
| Ohio | OH-TF1 | Miami Valley US&R |
| Pennsylvania | PA-TF1 | Commonwealth of Pennsylvania |
| Tennessee | TN-TF1 | Memphis Fire Dept. |
| Texas | TX-TF1 | State of Texas Urban Search & Rescue |
| Utah | UT-TF1 | Salt Lake Fire Dept. |
| Virginia | New York | Fairfax Co. Fire & Rescue Dept. |
| | VA-TF2 | Virginia Beach Fire Dept. |
| Washington | WA-TF1 | Puget Sound Task Force |

*Source:* FEMA, www.fema.gov

**Figure 4-2** New York, New York, September 27, 2001. FEMA, New York Fire Fighters, and the Urban Search and Rescue teams worked very closely throughout the cleanup effort at the World Trade Center. Photo by Bri Rodriguez/FEMA News Photo.

**Figure 4-3** Malibu, California, 1996. A California Department of Forestry official watches the wildfire as it burns up a hillside. FEMA News Photo.

## Other FEMA Response Resources

FEMA manages a cadre of nearly 4,000 temporary Disaster Assistances Employees (DAEs), who support FEMA response and recovery activities in the field in areas such as logistics, facility management, public affairs, community relations, and customer service. FEMA manages a mobile operations capability that provides communications and logistical support to state and local emergency officials.

### THE DAE EXPERIENCE

By Pat Glithero, FEMA Region V

Whenever the Federal Emergency Management Agency (FEMA) responds to a disaster, local and staff officials encounter and work with numerous FEMA staff, many of them women and men of FEMA's Disaster Assistance Employee (DAE) program. DAEs respond as needed to Presidential-declared emergencies and disasters across the nation and territories and remain until the Disaster Field Office (DFO) closes. They then return home to resume their lives as officials, businessmen, professionals, retirees, and in a myriad of other jobs.

DAEs join the cadre for many reasons. Some have benefited from FEMA programs in disasters and want to share their gratitude. Many believe strongly in programs designed to help fellow citizens. To many retirees, the DAE program allows a continued work opportunity alongside colleagues with whom they have worked for many years on a full-time basis. Some just appreciate working a job that uses their skills and allows them to feel they make a difference. Working in various geographic parts of the United States at a state/local level is a positive byproduct of disaster deployment. FEMA offers many opportunities for training classes and learning that might otherwise not be available in the private sector.

At the beginning, or end, of any DFO, though, what means most to many DAEs is the sense of camaraderie and family that DAEs share with each other and other FEMA staff. With a total workforce of less than 2,500 nationwide, FEMA DAEs and staff become part of a community that comes together as needed, does the job, and then parts officially until the next call.

*Source:* FEMA, www.fema.gov

### FEMA's MOBILE OPERATIONS CAPABILITY

Disasters may require resources beyond the capabilities of the local or state authorities. In response to regional requests for support, FEMA provides mobile telecommunications, operational support, life support, and power generation assets for the on-site management of disaster and all-hazard activities. This support is managed by the Response and Recovery Directorate's Mobile Operations Division (RR-MO).

The Mobile Operations Division has a small headquarters staff and five geographically dispersed Mobile Emergency Response Support (MERS) Detachments and the Mobile Air Transportable Telecommunications System (MATTS) to:

- Meet the needs of the government emergency managers in their efforts to save lives, protect property, and coordinate disaster and all-hazard operations.
- Provide prompt and rapid multimedia communications, information processing, logistics, and operational support to federal, state, and local agencies during catastrophic emergencies and disasters for government response and recovery operations.

The MERS and MATTS support the Disaster Field Facilities. They support the federal, state, and local responders—not the disaster victims.

## Available Support

Each of the MERS Detachments can concurrently support a large Disaster Field Office and multiple field operating sites within the disaster area. MERS is equipped with self-sustaining telecommunications, logistics, and operations support elements that can be driven or airlifted to the disaster location. MATTS and some of the MERS assets can be airlifted by C-130 military cargo aircraft.

The MERS and MATTS are available for immediate deployment. As required, equipment and personnel will deploy promptly and provide:

- Multimedia communications and information processing support, especially for the Communications Section, Emergency Support Function (ESF) #2 of the Federal Response Plan (FRP)
- Operational support, especially for the Information and Planning Section, ESF #5 of the FRP
- Liaison to the Federal Coordinating Officer (FCO)
- Logistics and life support for emergency responders
- Automated information and decision support capability
- Security (facility, equipment, and personnel) management and consultation

Most equipment is preloaded or installed on heavy-duty, multiwheel drive trucks. Some equipment is installed in transit cases.

*Source:* FEMA, www.fema.gov

# COMMUNICATIONS AMONG RESPONDING AGENCIES

## General

Overlapping responsibilities and unclear delineation make communications among responding agencies crucial. Responding agencies to a disaster event may include emergency management agencies from all levels of government, nongovernmental organizations, other responding government agencies, such as law enforcement and public health, the medical and scientific communities, and even businesses. Communications among these agencies is recognized as a current Achilles' heel in the emergency management field. Issues of authority and structure are difficult to resolve, and operations are often performed in an ad hoc fashion; however, improvement in this area is becoming a point of emphasis, and technological advancements are facilitating better communications as well.

The costs of poor coordination and communication can be high. A slow or ineffective initial disaster response disproportionately increases human losses. Also, poorly coordinated and perceived response efforts can damage political careers and the reputations of agencies. After Hurricane Andrew in Florida in 1992, it became apparent early in relief efforts that there were communications and coordination problems between FEMA, the state emergency management system, local agencies, and the governor's office. Many political analysts feel that the poor public perception of the response cost President George H.W. Bush votes in the 1992 Presidential election.

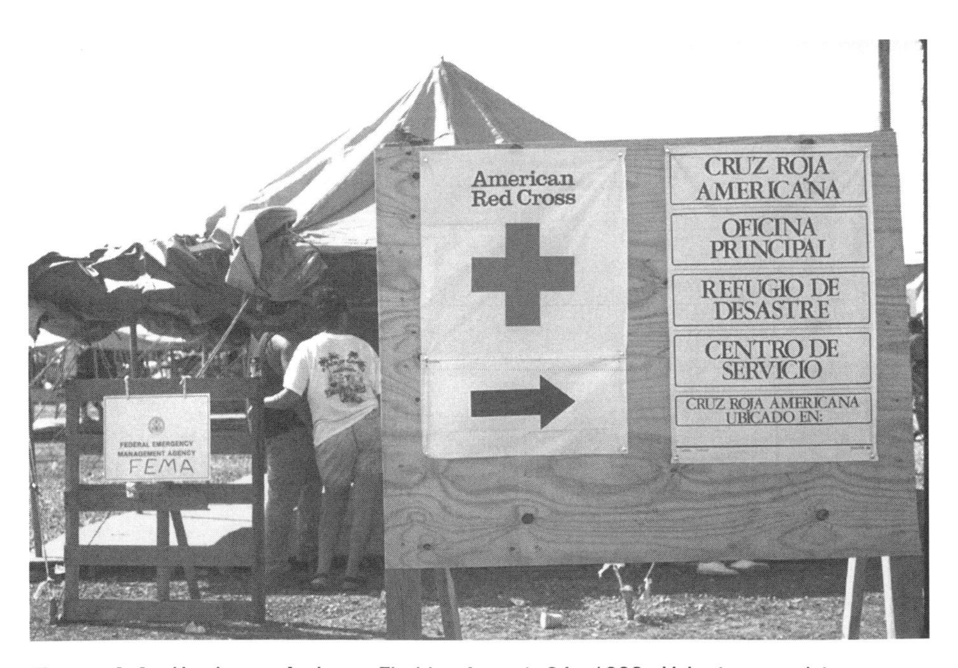


**Figure 4-4** Hurricane Andrew, Florida, August 24, 1992. Volunteer assistance was received from volunteer organizations, including the American Red Cross and Salvation Army. One million people were evacuated and 54 died in this hurricane. FEMA News Photo.

## Federal Response Plan

The FRP is the major coordination mechanism for the various responding federal agencies during a major disaster. Emergency Support Function (ESF) #5 within the FRP outlines responsibilities for information and planning. FEMA has the lead role for this activity but is supported by most other partner agencies in this respect. Individuals performing this function collect, analyze, process, and disseminate information about a disaster or emergency in order to facilitate the federal government assistance activities. The response is coordinated at the federal, field, regional, and headquarters levels. Daily information updates are provided to the various elements of the operation. The overall purpose of the function is to provide a central collection point where situation information can be compiled, analyzed, and prepared for use by decisions makers.

The FRP also includes a communications function (ESF #2), which basically deals with telecommunications infrastructure and technology. The lead agency for this function is the National Communications System. Its job is to ensure the provision of federal telecommunications support to federal, state, and local response efforts, and to serve as the planning point for use of national telecommunications assets and resources.

## FEMA Operations Center

The FEMA Operations Center (FOC) serves as the site of overall coordination and situation assessment operations for major disasters. It maintains a 24-hour capability to monitor all sources of information. Regional Operation Centers (ROCs) are the initial coordinating point for federal response efforts, however. The FEMA Director serves as the Federal Coordinating Officer (FCO) of the FOC, and assumes coordination responsibilities, working with the state coordinating officer and local officials.

## Joint Information Center

The Joint Information Center (JIC) is also a valuable tool for getting emergency management partners on the same page. In disasters of catastrophic or nationally significant proportions, a JIC is established to coordinate the dissemination of information about all disaster response and recovery programs. Public Affairs Officers (PAOs) representing all of the federal, state, local, and voluntary agencies providing response or recovery services are invited to co-locate and be a part of JIC operations. Interagency coordination is one of the central functions of the JIC, and teamwork is a key to implementing successful public information and media affairs programs. JICs involve coordination among the FCO, the lead state PAO, the congressional liaison, community relations and disaster assistance program managers, and other public agency PAOs.

## Command and Control versus Coordination

It is generally agreed that some type of mechanism is needed to facilitate coordination and communications among responding partners. What is not agreed upon is the structure of such a mechanism. The argument pits the clear, hierarchical "command and control" model against the more flexible, ad hoc "coordination" model.

The command and control model was adopted from the Incident Command System (ICS) used by fire departments across the United States and has clear lines of authority and responsibility. The coordination model is less rigid and more collaborative. In general, the coordination model is becoming more popular than the traditional command and control structure. For one thing, the new breed of emergency manager is typically more of a recovery coordinator than a field general. Also, command and control structures can sometimes hamper communications. The commanding organization may have a value system and technical language that is distinct from those of partner organizations or the victims. The coordination model takes this variability into account and focuses on providing an open communications forum. The coordination model is also often better for negotiating turf battles among agencies and nongovernmental organizations providing overlapping services.

## Technology

There are many examples of technology improving communications among partners. The use of the Internet as such a tool is an obvious trend. The city of Seattle recently integrated its Website into its emergency communications plan. The site

**Figure 4-5** West Palm Beach, Florida, October 1999. An inside view of Palm Beach County's state-of-the-art Emergency Operations Center, which enables local, state, and federal emergency management teams to coordinate interagency disaster response. Photo by Ty Harrington/FEMA News Photo.

now provides immediate access to information for members of other departments, such as police and transportation. The site also contains a database of press releases and space for "current news," which have been coordinated through an information control system. One lesson learned from Seattle's experience is to ensure that staff members who update the site are centrally located with emergency responders within the EOC.

Many communities are now using wireless systems to improve communications. The City of San Francisco recently developed a wireless voice and data communications system for its public safety agencies. The system overcomes the limited coverage of radio systems and the problem of various departments using incompatible systems. Mobile and portable radios are now in use at the city's fire, police, and emergency agencies. The Departments of Public Health, Public Works, Water, and the mayor's EOC also use the system. Officials indicate that it will go a long way toward helping the city handle the almost 4,000 emergency 911 calls it receives daily.

Wireless communications sometimes have their own limitations, however. After the terrorist attacks of September 11, 2001, cellular phone use overwhelmed wireless networks and prevented some local police and officials from making critical calls. In response, the White House plans to give emergency crews and government officials priority on the nation's cellular telephone system. Already in the United States, about 800 public institutions with emergency communications systems are given priority over regular users during an emergency. A similar effort is being initiated in Japan. After the 1995 Great Hanshin earthquake, incoming calls to Japan increased 50-fold and swamped the network.

## CONCLUSION

Responding to disaster events is the most visible activity that any federal, state, or local emergency management agency conducts. The politicians, the media, and the general public rate the success of an emergency management organization by how well it functions in the response phase of a disaster. A successful disaster response at any level of government requires a strong command and control system, clear lines of communications, and coordination of numerous agencies from multiple jurisdictions. Local first responders—fire, police, and emergency medical technicians—are on the scene first. Local and state emergency managers coordinate resources and assess the damage and the capacity of their jurisdictions to effectively respond. For major disaster events, a Presidential disaster declaration activates the FRP that delivers the full resources of the federal government in support of local and state authorities.

The key to the success of the emergency management system in the United States is the commitment of this country's elected officials to use the government to come to the aid of its citizens when a crisis occurs. The response process as described in this chapter ensures that government at all levels is capable of fulfilling this commitment.

# CASE STUDIES

## CASE STUDY: OKLAHOMA CITY

On April 19, 1995, an explosion rocked the federal plaza in Oklahoma City. The Alfred P. Murrah Federal Building was destroyed after a bomb, which was placed in a rental truck next to the building, was detonated. Upon arriving in the area, first responders witnessed smoke and fire coming from the Water Resource building. Believing that it was a natural gas explosion, it was not until EMS personnel entered this building that they noticed the gaping hole in the Murrah Building. The Fire Chief's first step was to have a single command center, which incorporated all buildings and victims within a one-mile radius. There were 33 fire stations, with at least 1,000 firefighters, 52 pieces of rescue apparatus that responded to the scene.

Within 45 minutes after notification from the Oklahoma Department of Civil Emergency Management, FEMA deployed staff to Oklahoma City. FEMA coordinated the federal response to the Oklahoma City bombing and later worked closely with state and local officials on recovery efforts. The President signed an Emergency Declaration within eight hours of the occurrence. This was the first time section 501(b) of the Stafford Act, granting FEMA the primary federal responsibility for responding to a domestic consequence management incident, was ever used. The President subsequently declared a major disaster on April 26, 1995. Because the disaster site was also a federal crime scene, FEMA appointed a liaison to the FBI to coordinate site access, support requirements, public information, and other issues. The coordinated work among federal agencies in Oklahoma City led to the further clarification of agency and department roles in crisis and consequence management.

Harsh lessons were learned in Oklahoma City. A situation arose when local radio stations requested that all medical personnel should respond to the disaster area. A nurse who answered the call was killed by falling debris while trying to rescue victims in the building. A term constantly used after the bombing was the Oklahoma Standard. Oklahoma had personnel on scene within 30 minutes. Federal officials were notified within minutes of the disaster. Volunteer services were immediate, and because this was a local disaster, everyone took responsibility to do whatever they could to help. Hospital personnel established an effective and efficient triage system. Phone numbers, Internet sites, and briefings were launched within hours of the disaster. The American Red Cross, as in all disasters, was quick to respond with personnel and supplies to help family members of those who were injured or killed in the bombing. The Salvation Army responded within hours with food and supplies. By the end of the day, the Salvation Army had deployed seven units to provide services to the workers and the victims. Law Enforcement and EMS personnel had up-to-date training. Oklahoma had excellent coordination with the Public Works Department, the National Weather Service, and the National Guard. The Department of Public Safety also had a predetermined disaster plan in place.

**Figure 4-6** Aurora, Illinois, July 1993. Illinois flood victim gets food from the Salvation Army. Photo by Liz Roll/FEMA News Photo.

Although there were some initial problems with communication, this was resolved within an hour as a result of support from Cellular One and Southwest Bell. They were able to clear lines, reconfigure their systems, and dispatch cell phones to personnel on scene. But most important was that the Oklahoma Highway Patrol could talk directly with personnel from federal agencies that were on the scene. A Department of Safety technician was able to program radios within 45 minutes of the disaster. Like most major cities, Oklahoma is equipped with 800MHz radios that can be linked with systems throughout the region. In any disaster, communication is the first line of defense in a successful response. It is essential that hospitals, rescue personnel, site commanders, and law enforcement officials have the ability to talk to one another. This was necessary to update the Disaster Field Office about the status of the response as well as to obtain needed personnel and supplies throughout the response. The only glitch was that the police were limited to those with whom they could communicate.

## CASE STUDY: HURRICANE FLOYD

On September 14, 1999, FEMA began mobilizing federal resources in preparation for possible landfall by Hurricane Floyd. While in previous years states had to wait for the disaster to strike before obtaining FEMA assistance, in the case of Hurricane Floyd, FEMA took a proactive stance by activating Emergency

Response Teams, allocating funds to local communities for law enforcement, and working with the Tropical Predication Center to monitor Hurricane Floyd's track. The ROC was put into action three days before the actual landfall of Hurricane Floyd.

On September 16, 1999, Hurricane Floyd made landfall near Cape Fear, North Carolina. The Category II hurricane had sustained winds of 110 miles per hour, but unlike Hurricane Andrew, the local first responders in coordination with FEMA were better prepared to handle this disaster. Emergency materials, generators, sheeting, tarp, bottled water, blankets, and clothing were identified and available for immediate delivery. Disaster Medical Assistance Teams (DMATs) had been placed on alert to provide medical services. Public works, including engineers, electricians, phone company employees, and public work personnel were also prepared for deployment to the area. Although forecasters thought that Floyd would hit Florida or Georgia, FEMA officials were mobile as the hurricane continued to track farther north. On September 15, 1999, President Clinton signed emergency declarations for North and South Carolina to fund law enforcement officials to help evacuate the areas. More than 2,100 employees were prepared to respond to the disaster. FEMA Urban Search and Rescue Teams from Indiana, Maryland, and Pennsylvania were activated. Upon the hurricane reaching land, FEMA's Mobile Emergency Response System (MERS) provided communication support to the affected communities.

FEMA's proactive response before landfall ensured that those affected by the hurricane would have the needed materials and services to help in the recovery phase. While the rain was still falling, FEMA established their toll-free service line. Within days, people were receiving financial aid to help them through the disaster. Although FEMA took some flack from certain areas of North Carolina and Virginia because of the long-lasting flooding, lives were saved and damage was reduced because of FEMA's and the 27 agencies' response to the hurricane.

## CASE STUDY: HURRICANE ANDREW

On August 24, 1992, Hurricane Andrew, a Category 4 hurricane, made landfall over Dade County, Florida. For everything that went right during the response for Hurricane Floyd, the opposite was true for Hurricane Andrew. "When Hurricane Andrew was approaching Florida and the advance element of the federal emergency response team deployed to the state emergency operations center in Tallahassee, it was evident that the state lacked sufficient space and resources to coordinate an operation to handle a disaster caused by a major hurricane like Andrew" (FEMA, 1993). In a postdisaster audit of FEMA's disaster management performance after Hurricane Andrew, the Inspector General noted that "state officials acknowledged that their initial assessment of requirements for federal assistance were too low, and that at first they were resistant to the idea of a massive flood of federal resources into south Florida" (FEMA, p. 41). Other problems noted by the Inspector General included a failure on the part of the state to request certain federal services because the state was reluctant to incur its

25 percent cost share and the lack of awareness of certain services by both state and local officials.

What became evident in the first weeks after Andrew was that FEMA and the overall federal response as well as the Florida response were uncoordinated, confused, and often inadequate (FEMA, 1993). FEMA requested its Inspector General to conduct a postdisaster audit, and Governor Chiles issued an executive order (92-242) establishing the Governor's Disaster Planning and Response Review Committee "to evaluate current state and local statutes, plans and programs for natural and man-made disasters, and to make recommendations to the Governor and the State Legislature" not later than January 15, 1993 (FEMA, 1993). The national emergency management system was acknowledged as being broken, and both the federal government and the state wanted to know why and what should be done to improve it.

The one key factor was that FEMA had yet to obtain clarification about its authority to supercede all other government agencies during a disaster. The Inspector General's report tasks each area that FEMA failed to perform. From preparation, to response and recovery, FEMA and federal officials dropped the ball. If it had not been for DOD intervention, people would have been left to their own device in seeking medical assistance, shelter, food, and water. Because federal agencies had to have a formal declaration declared, they were slow in responding and providing assistance to the people of Florida.

Without electricity, FEMA was unable to disseminate the needed information to the communities. Telephone lines, radio, and TV stations were disrupted for the first few days. People were not aware of the services available to them until days after the hurricane had struck. Although there was a FEMA employee at the Emergency Operation Center, he did not have the resources or the communication capabilities to get the needed response. The Defense Coordinating Officer (DCO) who was assigned to the Emergency Response Team was continually being drawn away from his assignment and also had his role continually expanded or changed during the response (FEMA, 1993).

The Inspector General's 200-page report took every aspect of the response and recovery phase into account and discussed in detail what needed to be done by local, state, and federal agencies for future catastrophic events. The report took into account the duplication of efforts by volunteer organizations and the lack of communication among the multiple federal agencies that had responded to Florida. Most consider President Bush's election loss to be partly attributable to the federal government's inability to manage domestic disasters.

With the Inspector General's report in hand, FEMA director James Witt moved forward on his goal to make FEMA the lead agency in emergency and disaster management. With the Federal Response Plan rewritten and clarification made, FEMA has successfully moved forward in using the FRP as a foundation that can be used during all disasters.

# 5. The Disciplines of Emergency Management: Recovery

## INTRODUCTION

There is often a theoretical debate over when the response function ends and the recovery function begins. For this book, the response function is classified as the immediate actions to save lives, protect property, and meet basic human needs. The recovery function is not so easily classified. This function often begins in the initial hours and days following a disaster event and can continue for months and in some cases years, depending on the severity of the event.

Unlike the response function, where all efforts have a singular focus, the recovery function or process is characterized by a complex set of issues and decisions that must be made by individuals and communities. Recovery involves decisions and actions relative to rebuilding homes, replacing property, resuming employment, restoring businesses, and permanently repairing and rebuilding infrastructure. The recovery process requires balancing the more immediate need to return the community to normalcy with the longer-term goal of reducing future vulnerability. The recovery process can provide individuals and communities with opportunities to become more economically secure and improve the overall safety and quality of life.

Because the recovery function has such long-lasting effects and usually high costs, the participants in the process are numerous. They include all levels of government, the business community, political leadership, community activists, and individuals. Each of these groups plays a role in determining how the recovery will progress. Some of these roles are regulatory, such as application of state or local building ordinances, and some, such as the insurance industry, provide financial support. The goal of an effective recovery is to bring all of the players together to plan, finance, and implement a recovery strategy that will rebuild the disaster-affected area safer and more secure as quickly as possible.

As noted in the previous chapter, the precipitating event for an area affected by a disaster is the Presidential declaration of disaster under the Stafford Act. Recovery activities begin immediately after a Presidential declaration as the agencies of the federal government collaborate with the state in the affected area in coordinating the implementation of recovery programs and the delivery of recovery services.

In the period of 1990 to 1999, FEMA spent more than $25.4 billion for declared disasters and emergencies compared to $3.9 billion in current dollars for 1980–1989. For the 1990–1999 period, more than $6.3 billion was provided in grants for temporary housing, home repairs, and other disaster-related needs for individuals and families. An additional $14.8 billion went to states and local governments for cleanup and restoration projects, including more than $1.37 billion for mission-assigned work undertaken by other federal agencies. In the 1990s, a total of 88 declarations were issued for hurricanes and typhoons, for which FEMA obligated more than $7.78 billion for disaster costs. The most costly to FEMA was Hurricane Georges in 1998, followed closely by Hurricane Andrew in 1992.

The most frequently declared disaster type was flooding resulting from severe storms, with more than $7.3 billion committed by FEMA for response and recovery costs. The most costly were the Midwest floods in 1993 and the Red River Valley Floods in 1997.

By December 2001, the disaster assistance provided by FEMA, the Small Business Administration (SBA), and the state of New York for the September 11, 2001, World Trade Center event had reached $700 million. Recovery costs for this disaster as of December 5, 2001, included the following:

- More than $344 million in public assistance funds to help New York City repair damaged infrastructure, restore critical services, and remove, transport, and sort debris.
- More than $196 million in individual assistance approved in the form of grants and loans. This assistance includes temporary disaster housing assistance, mortgage and rental assistance, disaster food stamps, individual and family grants, and SBA low-interest loans to homeowners and businesses.
- More than $151 million provided through other agencies, including the U.S. Army Corps of Engineers, Disaster Medical Assistance Teams from the Department of Health and Human Services, and FEMA's Urban Search-and-Rescue Task Force.

Without a doubt, the federal government plays the largest role in providing the technical and financial support for recovery. For that reason, this chapter focuses on the federal role in the disaster recovery function. It discusses the structure and the various programs available to assist individuals and communities in the postdisaster environment. The various national voluntary organizations that provide some assistance for recovery are briefly referenced, and several case studies are included to demonstrate the different types of recovery.

As noted earlier, the decisions during recovery are predominantly driven by local government. At the end of the chapter is a listing of potential planning tools for the recovery process. This, along with a more encompassing discussion of the complexities of, and roles and responsibilities of the various players in, recovery can be found in a book prepared for FEMA by the American Planning Association entitled *Planning for Post-Disaster Recovery and Reconstruction.*

## THE FEDERAL RESPONSE PLAN FOR DISASTER RECOVERY OPERATIONS

Issued in 1992, the Federal Response Plan (FRP) outlines how the federal government implements the Robert T. Stafford Disaster Relief and Emergency Assistance

Act, as amended, to assist state and local governments when a major disaster or emergency overwhelms their ability to respond effectively. The FRP describes the policies, planning assumptions, concept of operations, response and recovery actions, and responsibilities of 27 federal departments and agencies, including the American Red Cross, that guide federal operations following a Presidential declaration of a major disaster or emergency.

The FRP describes the structure for organizing, coordinating, and mobilizing federal resources for disaster recovery. The Stafford Act assigns to FEMA the principal coordination function—the interactive process by which multiple federal assistance programs are reviewed, initiated, implemented, and delivered to address the unique needs of a particular disaster area.

The fundamental assumption in the FRP is that recovery is a cooperative effort among federal, state, and local governments, voluntary agencies, and the private sector in partnership. A Federal Coordinating Officer (FCO) is appointed by the FEMA Director on behalf of the President and coordinates federal activities from the Disaster Field Office (DFO). The FCO works in partnership with the State Coordinating Officer (SCO), who is named by the governor. In conjunction with the SCO, the FCO determines the need for Disaster Recovery Centers (DRCs) in the disaster area. State and federal agencies staff the DRCs with knowledgeable officials who provide recovery program information, advice, counseling, and technical assistance. Voluntary organizations are encouraged to provide leadership and to coordinate with federal, state, and local governments in recovery planning and program implementation.

The practical work of implementing the recovery process occurs at the DRCs. Two organizational structures, or branches, divide the recovery assistance functions. These branches assess state and local recovery needs at the outset of the disaster and relevant time frames for program delivery. The Human Services Branch coordinates assistance programs to help individuals, families, and businesses meet basic needs and return to self-sufficiency (see Figure 5-1). It also handles the donations management function. The Infrastructure Support Branch coordinates assistance programs to aid state and local governments and eligible private nonprofit organizations to repair or replace damaged public facilities. The two branches assist in identifying appropriate agency assistance programs to meet applicant needs, synchronizing assistance delivery, and encouraging incorporation of mitigation measures where possible. In addition to the work of the DRCs, Applicant Briefings are conducted for local government officials and certain private nonprofit organizations to inform them of available recovery assistance and how to apply.

Federal disaster assistance available under a major disaster falls into three general categories: Individual Assistance, Public Assistance, and Hazard Mitigation Assistance. Individual Assistance is aid to individuals, families, and business owners. Public Assistance is aid to public and certain private nonprofit entities for emergency services and the repair or replacement of disaster-damaged public facilities. Hazard Mitigation Assistance is funding available for measures designed to reduce future losses to public and private property. A detailed description of the first two types of assistance follows. More information on Hazard Mitigation Assistance can be found in Chapter 3.

**Figure 5-1** Rocky Mount, North Carolina, September 29, 1999. A new life awaits residents whose homes were flooded by the rains from Hurricane Floyd. These manufactured homes, located near Rocky Mount, North Carolina, will house more than 300 families. Photo By Dave Saville/FEMA News Photo.

## FEMA'S INDIVIDUAL ASSISTANCE RECOVERY PROGRAMS

Individual Assistance programs are oriented to individuals, families, and small businesses, and the programs include temporary housing assistance, individual and family grants, disaster unemployment assistance, legal services, and crisis counseling. The disaster victim must first register for assistance and establish eligibility. Three national centers provide centralized disaster application services for disaster victims. FEMA's National Processing Service Centers (NPSCs) are located in Denton, Texas; Berryville, Virginia; and Hyattsville, Maryland.

Since the first national center opened in 1994, more than 2.5 million applications have been processed and 2.8 million calls taken for more than 275 major disasters. These NPSCs house an automated teleregistration service, through which disaster victims apply for Disaster Housing and the Individual and Family

Grant program, and through which their applications are processed and their questions answered.

This automated system provides automatic determination of eligibility for about 90 percent of Disaster Housing cases, usually within 10 days of application. The other 10 percent of cases, which may need documentation, take a little longer. Cases are also automatically referred to the state for possible grant assistance if the applicant's needs exceed the Disaster Housing program and the individual cannot qualify for a disaster loan from the Small Business Administration.

Following the September 11 events, FEMA was concerned that many individuals and businesses had not sought help in the aftermath of the attack. Working with the Advertising Council, and a volunteer ad agency, Muezzin Brown & Partners, a public service advertising campaign was developed to let viewers know that assistance was available by calling FEMA's toll-free registration number. The advertisements were distributed to electronic and media outlets in New York, New Jersey, Connecticut, Pennsylvania, and Massachusetts.

## Disaster Housing Program

The Disaster Housing Program assures that people whose homes are damaged by disaster have a safe place to live until repairs can be completed. These programs are designed to provide funds for expenses that are not covered by insurance and are available to homeowners and renters who are legal residents of the United States and who were displaced by the disaster (see Figure 5-2).

**Figure 5-2** South-facing Long Beach on Oak Island, North Carolina, September 17, 1999. Hurricane Floyd brought a devastating 15 feet of storm surge that damaged or destroyed hundreds of houses along this community's oceanfront and flattened its frontal sand dunes. Here, even elevation failed to save this home. Additional strapping, upgraded mitigation work, might have helped. Photo by Dave Gatley/FEMA News Photo.

- *Lodging expenses reimbursement* provides a check for reimbursement for the cost of short-term lodging such as hotel rooms, incurred because of damage to a home or an officially imposed prohibition against returning to a home.
- *Emergency minimal repair assistance* provides a check to help repair a home to a habitable condition.
- *Temporary rental assistance* provides a check to rent a place for the predisaster household to live.
- *Mortgage and rental assistance* provides a check to pay the rent or mortgage to prevent evictions or foreclosure. In order to qualify, the applicant must be living in the same house before and after the disaster and have a documented disaster-related financial hardship that can be verified by FEMA.

## Individual and Family Grant Program

The Individual and Family Grant (IFG) Program provides funds for the necessary expenses and serious needs of disaster victims that cannot be met through insurance or other forms of disaster assistance. The state administers the program and pays 25 percent of the grant amount, while the federal government provides the remaining 75 percent. The state also receives up to 5 percent of the federal share of the program for administrative costs. The maximum amount of a grant for each family or individual in fiscal year 2002 is $14,800, and the amount is adjusted annually for inflation.

The average grant, though, tends to be in the range of $2,000 to $4,000. Among the needs that can be met are housing, personal property, medical, dental, funeral, transportation, and required flood insurance premiums. Before an applicant can receive assistance for housing and personal property, the individual may be required to apply to the Small Business Administration (SBA) for a disaster loan. The SBA can provide three types of disaster loans to qualified homeowners and businesses to repair or replace homes, personal property, or businesses that sustained damages not covered by insurance:

- *Home disaster loans* provide funds to homeowners and renters to repair or replace disaster-related damages to home or personal property.
- *Business physical disaster loans* provide funds to business owners to repair or replace disaster-damaged property, including inventory and supplies.
- *Economic injury loans* provide capital to small businesses and to small agricultural cooperatives to assist them through the disaster recovery period. If the SBA determines that the individual is ineligible for a loan, or if the loan amount is insufficient to meet the individual's needs, then the applicant is referred to the IFG program.

IFG recipients who live in a Special Flood Hazard Area and receive assistance as a result of flood damages to their home and/or personal property will be provided with flood insurance coverage for three years under the National Flood Insurance Program (NFIP) group flood insurance policy. The three-year coverage is part of the grant award. The flood insurance must be kept active for the individual to be eligible to receive federal assistance for any future flood-related losses.

## Disaster Unemployment Assistance

The Disaster Unemployment Assistance (DUA) program provides unemployment benefits and reemployment services to individuals who have become unemployed because of major disasters, and who are not eligible for disaster benefits under regular unemployment insurance programs.

## Legal Services

The Young Lawyer's Division of the American Bar Association, through an agreement with FEMA, provides free legal assistance to low-income disaster victims. The assistance that the participating lawyers provide is for insurance claims; counseling on landlord/tenant problems; assistance in consumer protection matters, remedies, and procedures; and replacement of wills and other important legal documents destroyed in a major disaster. This assistance is intended for individuals who are unable to secure legal services adequate to meet their needs as a consequence of a major disaster.

## Special Tax Considerations

Taxpayers who have sustained a casualty loss from a declared disaster may deduct that loss on the federal income tax return for the year in which the casualty occurred or through an immediate amendment to the previous year's return. Businesses may file claims with the Bureau of Alcohol, Tobacco and Firearms (ATF) for payment of federal excise taxes paid on alcoholic beverages or tobacco products lost, rendered unmarketable, or condemned by a duly authorized official under various circumstances, including where a major disaster has been declared by the President.

## Crisis Counseling

The Crisis Counseling Assistance and Training Program is designed to provide short-term crisis counseling services to people affected by a presidentially declared disaster. The purpose of the crisis counseling is to help relieve any grieving, stress, or mental health problems caused or aggravated by the disaster or its aftermath. These short-term services are provided by FEMA as supplemental funds granted to state and local mental health agencies. The American Red Cross, the Salvation Army, and other voluntary agencies as well as churches and synagogues also offer crisis counseling services.

## Cora Brown Fund

Cora C. Brown of Kansas City, Missouri, died in 1977 and left a portion of her estate to the United States to be used as a special fund solely for the relief of human suffering caused by natural disasters. The funds are used to assist victims/survivors of presidentially declared major disasters for disaster-related needs that have not or will not be met by government agencies or other organizations.

# FEMA'S PUBLIC ASSISTANCE GRANT PROGRAMS

FEMA, under the authority of the Stafford Act, administers the Public Assistance Program. The Public Assistance Grant Program provides federal assistance to state and local governments and to certain private nonprofit (PNP) organizations. These grants allow them to recover from the impact of disasters and to implement mitigation measures to reduce the impacts from future disasters. The grants are aimed at governments and organizations with the final goal to help a community and its citizens recover from devastating major disasters. The federal share of assistance is not less than 75 percent of the eligible cost for emergency measures and permanent restoration. The state determines how the nonfederal share is split with the applicants.

Eligible applicants include the states, local governments, and any other political subdivision of the state, Native American tribes, Alaska Native Villages, and certain PNP organizations. Eligible PNP facilities include educational, utility, irrigation, emergency, medical, rehabilitation, temporary or permanent custodial care facilities, and other PNP facilities that are open to the public and provide essential services of a governmental nature to the general public. The work must be required as the result of the disaster, be located within the designated disaster area, and be the legal responsibility of the applicant. PNPs that provide critical services such as power, water, sewer, wastewater treatment, communications, or emergency medical care may apply directly to FEMA for a disaster grant. All other PNPs must first apply to the SBA for a disaster loan. If the loan is declined or does not cover all eligible damages, the applicant may reapply for FEMA assistance.

Work that is eligible for supplemental federal disaster grant assistance is classified as either emergency work or permanent work:

- *Emergency work* includes debris removal from public roads and rights-of-way as well as from private property when determined to be in the public interest. This may also include protective measures performed to eliminate or reduce immediate threats to the public.
- *Permanent work* is defined as work that is required to restore an eligible damaged facility to its predisaster design. This effort can range from minor repairs to replacement. Some categories for permanent work include roads, bridges, water control facilities, buildings, utility distribution systems, public parks, and recreational facilities. With extenuating circumstances the deadlines for emergency and permanent work may be extended.

As soon as possible after the disaster declaration, the state, assisted by FEMA, conducts the Applicant Briefings for state, local, and PNP officials to inform them of the assistance that is available and how to apply for it. A Request for Public Assistance must be filed with the state within 30 days after the area is designated eligible for assistance. A combined federal, state, and local team works together to design and deliver the appropriate recovery assistance for the communities. In determining the federal costs for the projects, private or public insurance can play a major role. For insurable buildings within special flood hazard areas (SFHAs) and damaged by floods, the disaster assistance is reduced by the amount of insurance settlement that would have been received if the building and its contents had been fully covered by a standard NFIP policy. For structures located outside of an SFHA, the amount is reduced by the actual or anticipated insurance proceeds.

In 1998, FEMA redesigned the Public Assistance program to provide money to applicants more quickly and to make the application process easier. The redesigned program was approved for implementation on disasters declared after October 1, 1998. This redesigned program placed new emphasis on people, policy, process, and performance. The focus of the program was also modified to provide a higher level of customer service for disaster recovery applicants and to change the role of FEMA from inspection and enforcement to an advisory and supportive role.

## OTHER FEDERAL AGENCY DISASTER RECOVERY FUNDING

Other federal agencies have programs that contribute to social and economic recovery. Most of these additional programs are triggered by a Presidential declaration of a major disaster or emergency under the Stafford Act; however, the Secretary of Agriculture and the Administrator of the SBA have specific authority relevant to their constituencies to declare a disaster and provide disaster recovery assistance. All of the agencies are part of the structure of the FRP. This section does not provide a complete list of all disaster recovery programs available after a disaster declaration, but provides a summary of many of the federal agencies in addition to FEMA that provide disaster recovery programs. These agencies include the following:

- U.S. Army Corps of Engineers
- Department of Housing and Urban Development
- Small Business Administration
- U.S. Department of Agriculture
- Department of Health and Human Services
- Department of Transportation
- Department of Commerce
- Department of Labor

A more comprehensive list is available in the Catalog of Federal Domestic Assistance (CFDA), available through the Federal Assistance Programs Retrieval System. Each automated edition is revised in June and December (see Figure 5-3).

### U.S. Army Corps of Engineers

In a typical year, the Corps of Engineers responds to more than 30 Presidential disaster declarations, plus numerous state and local emergencies. Under the FRP,

**Figure 5-3** New York, New York, October 30, 2001. FEMA/NY State Disaster Field Office personnel meet to coordinate federal, state, and local disaster assistance programs. Photo by Andrea Booher/FEMA News Photo.

the Army has the lead responsibility for public works and engineering missions. For example, after the events of September 11, 2001, the Army provided technical assistance for the debris removal operation. By December 2001, more than 661,430 tons of debris had been moved to the Staten Island landfill.

## Department of Housing and Urban Development

The Department of Housing and Urban Development (HUD) provides flexible grants to help cities, counties, and states to recover from presidentially declared disasters, especially in low-income areas, subject to availability of supplemental appropriations. When disasters occur, Congress may appropriate additional funding for the Community Development Block Grant (CDBG) and HOME programs to rebuild the affected areas and bring crucial seed money to start the recovery process. Because it can fund a broader range of recovery activities than most other programs, CDBG disaster recovery assistance supplements recovery assistance from FEMA and helps communities and neighborhoods that otherwise might not recover because of limited resources.

The CDBG program funds have been especially useful to communities that are interested in incorporating mitigation into their recovery process. These funds have been combined with FEMA assistance to remove or elevate structures from the flood plain and to relocate residents and businesses to safer areas.

The HOME Program helps expand the supply of decent, affordable housing for low- and very low-income families by providing grants to states and local governments. Funds can be used for acquisition, new construction, rehabilitation, and tenant-based rental assistance. HOME disaster recovery grants are an important resource for providing affordable housing to disaster victims.

## Small Business Administration

The SBA Disaster Loan Program offers low-interest loans to assist in long-term recovery efforts for those who are trying to rebuild their homes and businesses in the aftermath of a disaster. Disaster loans from SBA help homeowners, renters, businesses of all sizes, and nonprofit organizations fund rebuilding efforts. The SBA Disaster Loan Program reduces federal disaster costs compared to other forms of assistance, such as grants, because the loans are repaid to the U.S. Treasury. The SBA can only approve loans to applicants who have a reasonable ability to repay the loan and other obligations from earnings. The terms of each loan are established in accordance with each borrower's ability to repay. Generally, more than 90 percent of the SBA's disaster loans are made to borrowers without credit available elsewhere and have an interest rate of around 4 percent. The disaster loans require borrowers to maintain appropriate hazard and flood insurance coverage, thereby reducing the need for future disaster assistance.

The SBA is authorized by the Small Business Act to make two types of disaster loans: physical disaster loans and economic injury disaster loans. Physical disaster loans are a primary source of funding for permanent rebuilding and replacement of uninsured disaster damages to privately owned real and/or personal property. Economic injury disaster loans provide necessary working capital until normal operations resume after a physical disaster.

In 2000, the SBA approved 28,218 loans for $1.028 billion. Since the inception of the program in 1953, the SBA has approved more than 1.5 million disaster loans for more than $28.5 billion. In 2001 after the September 11 events, the SBA approved more than $161 million in low-interest loans to more than 2,000 applicants for home repairs, business loans, and loans to assist small businesses suffering economic injury as a result of losses caused by the disaster.

## U.S. Department of Agriculture

The U.S. Department of Agriculture (USDA) Farm Service Agency (FSA) provides low-interest loan assistance to eligible farmers and ranchers to help cover production and physical losses in counties declared as disaster areas by the President or designated by the Secretary of Agriculture. The emergency loans can be used to restore or replace essential physical property, pay all or part of production costs associated with the disaster year, pay essential family living expenses, reorganize the farming operation, and refinance debts.

## Department of Health and Human Services

The Department of Health and Human Services (DHHS) is the lead federal agency responsible for implementing the health and medical portion of the FRP. Their activities provide support to individuals and communities affected by disasters, state and local mental health administrators, and other groups that respond to those affected by human-caused disasters (such as school violence). The Center for Mental Health Services (CMHS) within the DHHS works with FEMA to implement the Crisis Counseling Assistance and Training Program discussed earlier in this chapter.

The DHHS also provides disaster assistance for older Americans through its Administration on Aging (AoA). Older people often have difficulty obtaining necessary assistance because of progressive physical and mental impairments and other frailties that often accompany aging. Many older people, who live on limited incomes, and are sometimes alone, find it impossible to recover from disasters without special federal assistance service. The AoA's national aging network assists older persons by providing critical support such as meals and transportation, information about temporary housing, and other important services on which older adults often rely.

## Department of Transportation

Congress authorized a special program from the Highway Trust Fund for the repair or reconstruction of federal-aid highways and roads on federal lands that have suffered serious damage as a result of natural disasters or catastrophic failures from an external cause. The Department of Transportation (DOT) Federal Highway Administration (FHWA) administers the Emergency Relief Program, which supplements the commitment of resources by states, their political subdivisions, or other federal agencies to help pay for damages resulting from disasters. The applicability of the program to a natural disaster is based on the extent and intensity of the disaster.

## Department of Commerce

Within the Department of Commerce, the Economic Development Administration (EDA) administers programs and provides grants for infrastructure development, business incentives, and other forms of assistance designed to help communities alleviate conditions of substantial and persistent unemployment in economically distressed areas and regions. The EDA provides postdisaster economic assistance for communities affected by declared natural disasters. Funding for this program has been a problem over the years.

## Department of Labor

The Department of Labor (DOL) Disaster Unemployment Assistance (DUA) Program provides financial assistance to individuals whose employment or self-employment has been lost or interrupted as a direct result of a major disaster and who are not eligible for regular state unemployment insurance. Funding for this program comes from FEMA. The DUA is administered by the state agency responsible for providing state unemployment insurance.

The Workforce Investment Act of 1998 authorizes the U.S. Secretary of Labor to award National Emergency Grants to assist any state that has suffered an emergency or major disaster to provide disaster relief employment. These funds can be used to finance the creation of temporary jobs for workers dislocated by disasters to clean up and recover from the disaster, and to provide employment assistance to dislocated workers. Interestingly, in creating this program, Congress expanded eligibility beyond people affected by the disaster to dislocated workers and certain civilian Department of Defense employees affected by downsizing and certain recently separated members of the armed forces.

# NATIONAL VOLUNTARY RELIEF ORGANIZATIONS

National Voluntary Organizations Active in Disaster (NVOAD) coordinates planning efforts by many voluntary organizations responding to disaster in order to provide more effective service to people affected by disaster. Members include 34 national voluntary organizations active in disaster mitigation and response, 52 state and territorial chapters (VOADs), and dozens of local organizations. Once a disaster occurs, NVOAD or an affiliated state VOAD encourages members and other voluntary agencies to convene on site. The member organizations provide a wide variety of disaster relief services, including emergency distribution services, mass feeding, disaster child care, mass or individual shelter, comfort kits, supplementary medical care, cleaning supplies, emergency communications, stress management services, disaster assessment, advocacy for disaster victims, building or repair of homes, debris removal, mitigation, burn services, guidance in managing spontaneous volunteers, and victim and supply transportation. NVOAD maintains a close relationship with FEMA and encourages the state and local affiliates to work closely with the state and local emergency management agencies.

## The American Red Cross

Although the American Red Cross is not a government agency, its authority to provide disaster relief was formalized when, in 1905, the Red Cross was chartered by Congress to "carry on a system of national and international relief in time of peace and apply the same in mitigating the sufferings caused by pestilence, famine, fire, floods, and other great national calamities, and to devise and carry on measures for preventing the same." Red Cross disaster relief focuses on meeting people's immediate emergency disaster-caused needs and provides disaster assistance to individuals to enable them to resume their normal daily activities independently. The Red Cross provides shelter, food, and health and mental health services to address basic human needs. The Red Cross also feeds emergency workers, handles inquiries from concerned family members outside the disaster area, provides blood and blood products to disaster victims, and helps those affected by disaster to access other available resources.

The Red Cross is one of the nongovernmental organizations mentioned in the FRP and is the primary agency for ESF #6, Mass Care. The Red Cross functions as a federal agency in coordinating the use of federal mass care resources in a presidentially declared disaster or emergency, and coordinates federal assistance in support of state and local efforts to meet the mass care needs of victims of a disaster. This federal assistance supports the delivery of mass care services of shelter, feeding, and emergency first aid to disaster victims; the establishment of systems to provide bulk distribution of emergency relief supplies to disaster victims; and the collection of information to operate a Disaster Welfare Information system to report victim status and assist in family reunification.

## RECOVERY PLANNING TOOLS

Despite the pressures on politicians and community leaders to return to a period of normalcy as quickly as possible and because of federal incentives, public interest, and insurance retractions, more and more communities are looking at ways to reduce their future vulnerability. As disasters repeat themselves and the public sees the emotional and financial benefits of mitigation, communities are making the long-term investment in mitigation. For example, the devastating 1993 Midwest floods that occurred again in some areas in 1995 had a minimal impact in those towns where buyout and relocation programs were undertaken after the 1993 flood. The following is a partial list of policy areas and tools that should be considered by decision makers as they develop their recovery plan:

- *Land-use planning techniques*, including acquisition, easements, annexation, stormwater management, and environmental reviews
- *Zoning*, including special-use permits, historic preservation, setbacks, density controls, wetlands protection, floodplain, and coastal zone management
- *Building codes*, including design controls, design review, height and type, and special study areas (soil stability ratings)
- *Financial*, including special districts, tax exemptions, special bonds, development rights, property transfer, or use change fees
- *Information and oversight*, including public awareness and education, regional approaches and agreements, global information systems, town hall meetings, and public hearings

## CONCLUSION

As this chapter demonstrates, the federal government plays a significant role in initiating and funding the disaster recovery process. But for recovery to be effective, the planning and decision making must be done at the local level. With a disaster comes disruption and tragedy, but in the aftermath comes opportunity. Changes to FEMA's Stafford Act now require communities and states to have mitigation plans approved before the disaster. These plans, developed in the calm before an event happens, can become the blueprint for facilitating recovery and making communities less vulnerable in the postdisaster environment. Communities should strive to integrate preevent recovery and mitigation planning into their ongoing planning efforts. Such integration will allow for the political process to work, to include citizen participation, and to garner support for changes that will make their communities safer and more secure.

## CASE STUDIES

### CASE STUDY: FEDERAL ACTION PLAN FOR THE RED RIVER VALLEY FLOODS

In April 1997, the Red River flooded its banks, displacing more than 60,000 people and affecting communities especially hard in Grand Forks, North Dakota, and East Grand Forks, Minnesota. On April 7 and 8, a Presidential disaster was declared for severe spring storm conditions in North Dakota, South Dakota, and Minnesota. Following an April 22 visit to these communities, the President announced that all emergency measures and costs of debris removal under the Stafford Act would be covered 100 percent by the federal government so that the state and local governments could concentrate their resources on response and recovery efforts. The President also announced the formation of an interagency task force to develop a long-term recovery plan for the affected states. James Lee Witt, the director of FEMA at that time, chaired the effort.

The Federal Action Plan for Recovery identified three priorities for federal long-term recovery efforts: mitigation of flood hazards, housing, and reestablishing community sustainability. In conjunction with state and local governments, the action plan detailed a wide range of grants, loans, and technical assistance that the federal government would provide to ensure that community recovery needs were addressed. The President also ordered several federal departments to implement efforts to make the communities more disaster resistant. He directed the U.S. Army Corps of Engineers to aggressively pursue the development and implementation of structural and nonstructural flood protection works for the cities of Grand Forks and East Grand Forks.

FEMA and HUD were directed, in partnership with the states, to implement an accelerated program to purchase flood-damaged residences in the most severely devastated areas. FEMA, HUD, the Army Corps, the Economic Development Administration (EDA), and the SBA were directed to use all available authorities to support state and local rebuilding efforts and to incorporate mitigation to make the communities disaster resistant. The President also asked the affected communities to vigorously pursue mitigation and to manage development wisely to avoid future flooding events. He encouraged the residents of these communities to purchase and maintain flood insurance.

To address the issue of immediate and long-term housing availability and to maintain community continuity during the recovery process, the President directed FEMA to continue providing temporary housing on an expedited basis by providing emergency home repair grants, travel trailers to be sited next to unlivable damaged residences under repair, mobile homes for those facing longer-term displacement, and rental assistance. HUD, the Department of Commerce, the EDA, the USDA, and the SBA were directed to establish a recovery office in Grand Forks to help the communities create new housing resources through planning and design assistance, infrastructure funding, and continued low-interest loans to homes and businesses.

The Recovery Action Plan also addressed the challenge of reestablishing the sustainability of the community through preserving historic downtown and residential areas, attracting and retaining a workforce, building and repairing infrastructure and housing, and capitalizing small businesses. To help meet these challenges, the President directed HUD, the EDA, the SBA, the Army Corps, the USDA, FEMA, and the Department of Energy (DOE) to provide short-term and long-term planning and technical assistance to the communities most impacted. The SBA, HUD, the EDA, the USDA, and FEMA were directed to continue to make low-interest loans and targeted grants to support development of new business facilities, assist in relocation of businesses away from highly hazardous areas, stimulate private-sector investment, and address reestablishment and relocation of critical facilities, including water treatment plants. HUD, the SBA, the USDA, and the EDA were also directed to actively seek innovative solutions to the short-term capitalization of businesses, in particular small businesses.

The President directed FEMA to provide temporary classroom and administration facilities for schools and to support the communities' efforts in the design and siting of new construction of schools away from high-risk areas. FEMA was also directed to continue the repair, restoration, and mitigation of damaged infrastructure, including roads, bridges, hospitals, and other public and private nonprofit facilities.

Other agencies also helped address the immediate disaster recovery needs of these three states after the floods. The Department of Labor made nearly $10 million available under the Job Training Partnership Act Title III program to provide temporary jobs for disaster-affected workers in the three states. The Centers for Disease Control and Prevention (CDC) from the DHHS provided emergency assistance to the affected areas on environmental health, disease and injury surveillance, worker safety, and water quality. The Federal Highway Administration (FHA) allocated emergency funds to repair highways. The Environmental Protection Agency (EPA) provided technical assistance to the states on solid waste, pesticides, household hazardous waste, air monitoring, and underground storage tank issues.

In regard to these actions, James Lee Witt stated: "The Long Term Recovery Task Force developed recommendations that transcend our usual disaster programs. This innovative effort demonstrates the federal government's true commitment to the long-term recovery of communities in the three states deluged by the Red River of the North. In addition to helping these communities recover, we are committed to assisting state and local governments with the task of rebuilding safer and smarter to reduce future flood risks."

## CASE STUDY: LONG-TERM RECOVERY ACTION PLAN FOR HURRICANE GEORGES

On September 21, 1998, Hurricane Georges, sustaining winds as high as 150 miles per hour, struck Puerto Rico and dumped more than two feet of rain on the island. More than 100,000 residences were damaged or destroyed, and 31,500 people were forced to seek refuge in shelters. This was the worst natural disaster to hit Puerto Rico in 70 years, and a major disaster was declared for all 78 of Puerto Rico's municipalities. In response to the severity and scope of the destruction, the President activated the Long-Term Recovery Task Force composed of 15 federal departments, agencies, and offices, and headed by then FEMA director James Lee Witt. The President directed the group to develop an action plan to facilitate the coordination and delivery of federal recovery assistance to Puerto Rico.

The purpose of the Task Force is to coordinate and target the diverse disaster programs of more than a dozen federal agencies to ensure the greatest level of effective federal support. The Task Force worked in collaboration with representatives of the Government of Puerto Rico to identify five long-term recovery priorities: mitigation, housing, economic revitalization and sustainability, energy, and transportation.

The Government of Puerto Rico identified **mitigation** as one of the core elements of its vision for long-term recovery. Federal mitigation actions emphasized three areas: building codes, planning and coordination, and floodplain management. FEMA provided technical assistance for developing long-term strategies to reduce losses in future disasters and provided funding under the Hazard Mitigation Grant Program. The federal government also worked with Puerto Rico to acquire property and elevate structures in the floodplain. The U.S. Army Corps of Engineers worked with Puerto Rico to identify funding for and expedite construction of flood control projects.

Federal assistance for **housing** focused on repairing existing homes, addressing long-term shelter needs, replacing destroyed homes, restoring public housing, and providing technical assistance and training. FEMA provided funding assistance under the Disaster Housing Assistance program and the Individual and Family Grant program. Additional funding was provided through the SBA Home Disaster Loans and the USDA Rural Housing Service. HUD provided disaster funds through the Community Development Block Grant program. FEMA collaborated with Puerto Rico on improved housing design plans for low-income residents and also provided technical assistance and funding for the development of long-term sheltering options.

The federal government worked with Puerto Rico to put in place improvements to achieve the long-term benefits of **economic revitalization and sustainability**. In the agricultural sector this was accomplished through financial assistance for crop and physical losses, expanding agricultural insurance and coverage, and financial and technical assistance for conservation measures to reduce flooding and erosion. The USDA Risk Management Agency provided funding for crop loss insurance claims. The USDA Natural Resources Conservation Service provided financial and technical assistance to address flooding and soil erosion problems.

In the nonagricultural sector, the federal government provided community development planning assistance, supported small business recovery, encouraged new investment, proposed fiscal assistance, provided unemployment assistance, and promoted flood insurance for homeowners, renters, and businesses. HUD made available technical assistance for economic development strategies and financial packaging. The EDA provided a community planning grant to the University of Puerto Rico's Economic Development University Center and committed funds to Puerto Rico's Economic Development Bank for a revolving loan fund assistance program. The DOL provided funding to create temporary jobs to assist in the immediate and long-term cleanup and recovery efforts. The DOL also provided unemployment assistance.

Hurricane Georges caused 100 percent of the electrical service in Puerto Rico to be disrupted. Its failure crippled other basic services such as water and sewage treatment, telephone service, transportation, and local commerce. Federal assistance for the **energy** sector included providing resources for repairing electrical transmission and distribution lines, and recommendations for design improvements; emergency generators; and assistance for developing a more reliable electrical system. The cost for repairing the island's electrical system was paid by a combination of Puerto Rico's self-insurance coverage and funding through FEMA's public assistance program. Electric utility workers, trucks, and equipment were flown to the island to assist local crews. Emergency generators were provided to keep critical facilities operational, and plans were developed to keep some of the generators in place to provide backup power during future disasters. The Department of Energy, FEMA, and Puerto Rico examined mitigation measures to improve the disaster resistance of the electrical system through enhanced generation/transmission relationships, better power line placement, and placing poles deeper in the ground.

Key **transportation** issues that were addressed included repairing damaged roads and bridges, developing a reliable power source for the Tren Urbano project, and dredging harbors. The Army Corps removed tons of debris from roadways, installed four temporary bridges, and provided financial assistance for critical dredging activities to maintain safe harbor channels. The FHA and FEMA provided financial assistance for rebuilding the island's damaged transportation system. Mitigation measures were incorporated into road and bridge repairs to reduce the risk of such severe damage in the future. The Federal Transit Authority and FEMA worked with the Government of Puerto Rico to explore funding options to establish a reliable power source for the Tren Urbano, a San Juan metro-area mass transit system.

The Governor of Puerto Rico, Pedro Rosselló, stated that: "From the President on down, the federal government mobilized all of the resources at its disposal—even before the hurricane struck—and has earned the eternal gratitude of Puerto Rico's 3.9 million people for its role in helping us cope with this catastrophe. The scope of the response is illustrated by the fact that the President's Long-Term Recovery Task Force is rarely activated."

## CASE STUDY: UNIVERSITY OF HOUSTON O'QUINN LAW LIBRARY

Tropical Storm Allison formed on Wednesday evening, June 6, 2001, in the Gulf of Mexico southeast of Galveston, Texas, and eventually exited the United States on Sunday night, June 17, after passing through Florida and proceeding up the East Coast. Allison proved to be the most destructive tropical storm in U.S. history, costing 43 lives and nearly $5 billion. The storm hit Houston, Texas, especially hard, dumping between 30 to 40 inches of rain and causing an estimated $1 billion in damage. On June 9, 2001, President Bush declared a major disaster for the state of Texas, with 28 counties eligible for public assistance. The University of Houston O'Quinn Law Library was flooded with 8 feet of water after the heavy rains from Tropical Storm Allison.

The lower floor of the library filled nearly to the 12-foot ceilings with a mixture of water, oil, asbestos, and other pollutants. The 35,000 square feet of space in the lower level were equal to nearly two floors of a typical downtown skyscraper. The metal shelves were destroyed, partly by the tremendous weight of waterlogged books and partly by being literally exploded as the wet books began swelling and exerting tremendous sideways pressure. The library lost between 200,000 and 500,000 books, and damages were estimated at $30 million.

Through the Public Assistance program, FEMA approved $21.4 million for the replacement of 174,000 copies of law books and microfiche storage collection. The funding approved by FEMA was for two separate projects: one project in the amount of $1,204,600 was for the microfiche collection, and the other project in the amount of $27,295,196 was for law book replacement. FEMA provided 75 percent of the cost, with the remaining 25 percent coming from local sources. "With the support of all our communities, and major assistance from FEMA, not only have we recovered, but we're putting in place an even stronger and more secure resource for our law center faculty and students as well as the community," said University of Houston President, Arthur K. Smith.

# 6. The Disciplines of Emergency Management: Preparedness

## INTRODUCTION

Preparedness within the field of emergency management can best be defined as a state of readiness to respond to a disaster, crisis, or any other type of emergency situation. Preparedness is not only a state of readiness, but also a theme throughout most aspects of emergency management. If you look back into the history of the United States, you see the predecessors of today's emergency managers focusing on preparedness. The fallout shelters of the 1950s and the air raid wardens were promoting preparedness for a potential nuclear attack from the Soviet Union. An early 1970s study prepared by the National Governor's Association talked about the importance of preparedness as the first step in emergency management.

After the Three Mile Island Nuclear Power Plant incident in 1979, preparedness around commercial nuclear power plants became a major issue for continued licensing of these plants. The increased emphasis on preparing the public for a potential event through planning and education and preparing local responders through required exercises caused an increased focus on preparedness. The Nuclear Regulatory Commission's licensing requirements required local emergency plans, exercise of those plans, and evaluation of the exercises.

This process had a profound impact on the discipline of emergency management. This off-site preparedness planning process became the model for future emergency response plans. The required exercises were some of the first such activities. They brought a legitimacy and level of public and political exposure to the emergency management profession. Most people agree that the radiological emergency preparedness program that was initiated in the aftermath of Three Mile Island and that became part of the newly created federal agency, the Federal Emergency Management Agency (FEMA), was the start of modern emergency management discipline.

Since that time, preparedness has advanced significantly, and its role as a building block of emergency management continues. No emergency management organization can function without a strong preparedness capability. This capability is built through planning, training, and exercising. Preparedness activities have led toward an increased professionalism within the discipline of emergency management. Throughout the 1990s, FEMA was focused on supporting and enhancing these efforts, not just at the federal level but also throughout government and into the private sector.

All organizations in private, public, and government sectors are susceptible to the consequences of a disaster and must consider preparedness. For example, preparedness focuses not only on getting essential government services, such as utilities and emergency services, functioning at predisaster levels, but also on assisting businesses in quickly reopening to the public. Both of these key functions of preparedness help minimize the required time for the affected population to return to predisaster life. Business contingency planning has emerged as a profitable offshoot of government preparedness efforts.

This chapter discusses the preparedness cycle from a systems approach, preparedness programs, hazard preparedness, training programs, and exercise programs. The focus is on federal efforts, predominantly FEMA, and best practices are highlighted through several case studies.

## PREPAREDNESS: THE BUILDING BLOCK

Within the Federal Response Plan, there are 12 emergency functions, each of which relies on a level of preparedness. These functions are defined as transportation, communications, public works and engineering, firefighting, information and planning, mass care, resource support, health and medical services, urban search and rescue, hazardous materials, food, and energy. Each individual functional area must ensure its own preparedness in order to establish a systemwide posture that is ready to respond and act in an emergency.

All 12 functions are highly dependent on each other. For example, the functions of emergency communications must be prepared to establish emergency telecommunications support in order for the firefighters, who must be prepared with the equipment and training to extinguish the fires, to know where to go and coordinate with the urban search-and-rescue teams that locate and rescue victims, each of which must be provided timely transportation to reach the disaster scene.

Preparedness is therefore defined more fully by FEMA as

> the leadership, training, readiness and exercise support, and technical and financial assistance to strengthen citizens, communities, State, local and Tribal governments, and professional emergency workers as they prepare for disasters, mitigate the effects of disasters, respond to community needs after a disaster, and launch effective recovery efforts. (www.fema.gov)

## MITIGATION VERSUS PREPAREDNESS

Preparedness has been defined, and it has been mentioned that preparedness encompasses various aspects of response, but how does mitigation play into the equation? Mitigation is the cornerstone of emergency management. It's the ongoing effort to lessen the impact disasters have on people and property. Mitigation involves keeping homes away from floodplains, engineering bridges to withstand earthquakes, creating and enforcing effective building codes to protect property from hurricanes—and more.

Preparedness deals with the functional aspects of emergency management such as the response to and recovery from a disaster, whereas mitigation attempts to lessen these effects through predisaster actions—as simple as striving to create "disaster-resistant" communities.

## A SYSTEMS APPROACH: THE PREPAREDNESS CYCLE

As an academic field, as well as an applied practice in the public and private sector, emergency management has just recently been established. For this reason, it has thus far drawn on the fields of emergency medicine, fire suppression, and law enforcement for many of its foundations. Although these are tried-and-tested specialties, they also are steeped in tradition—relying less on academic or analytic processes. Without a foundation that ties academia and structured analytic methodologies with tradition, the extreme complexity of emergency management, often requiring coordination among tens to hundreds of individual agencies and organizations, will not be effectively managed. Therefore, a systematic approach must be established for emergency management as a whole, and specifically the steps necessary to reach preparedness.

The diagram depicted in Figure 6-1, used in terrorism planning, depicts the planning process, beginning with assessing the threats to a jurisdiction or business, be it natural or manmade, and working in a systematic approach toward a cyclical process to establish preparedness. This systematic and cyclical approach is defined by the continual evolution of the phases on the exterior ring—assessment, planning, preparation, and evaluation.

In this depiction, the interior ring defines each of the steps that organizations must work toward in order to be prepared. The first step is to identify what types of disasters, or threats, a jurisdiction, business, or any entity faces. Next, assessing the current vulnerability, or level of preparedness, will lead toward determining the

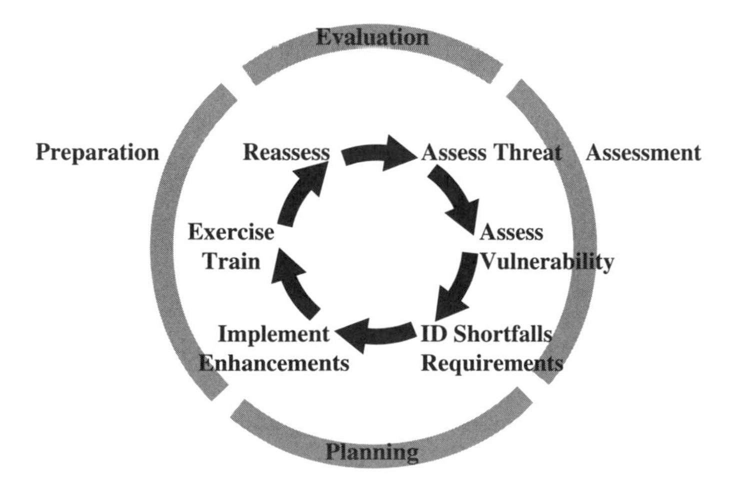


**Figure 6-1** Preparedness Planning Cycle.

shortfalls between current preparedness and the requirements to meet an improved preparedness posture. This improved posture may be determined through industrial standards set forth by organizations such as the National Fire Protection Association, who sets fire safety standards, or the International Organization for Standardization (ISO), one of the largest developers of standards and certifications. Local, state, and/or federal laws can also statutorily define a required level of preparedness.

Implementing enhancements or revamping complete systems then addresses these identified shortfalls. Exercises and training can then be used to test whether the enhancements or new systems are, in fact, meeting the standards determined in earlier stages. If they are, then the end goal of readiness or preparedness regarding a particular threat such as terrorism or floods is met.

The cyclical nature of this system is the fundamental aspect of the successive steps to be taken after determining whether a jurisdiction, or any type of entity, is prepared. Whether or not those standards are met, the entity must reexamine its threats because both natural and technological threats change constantly. The important realization that preparedness is a dynamic state and can either improve or diminish in a short time must be understood by the emergency management professional. Using the systems approach will ensure that an overall emergency management system is prepared, but more important, each of the individual functional areas is also prepared.

The importance and diversity of this vital aspect of preparedness planning can be demonstrated through the other types of assessment processes available. Another example available to emergency managers is provided by FEMA in its Capability Assessment for Readiness (CAR) program.

## FEMA's CAPABILITY ASSESSMENT FOR READINESS PROGRAM

FEMA and the National Emergency Management Association (NEMA) have joined together in partnership to develop an emergency management readiness and capability assessment system for state and local emergency managers. The result of this effort is the Capability Assessment for Readiness (CAR). This initiative further strengthens the Emergency Management Performance Grant (EMPG) Program that provides federal financial assistance to state and local governments. Also imbedded in the CAR are those important ingredients developed by the National Fire Protection Association (NPFA 1600) termed "Emergency Management Standards." The CAR process also provides for the assessment component of the EMPG process that will continue to evolve in coming years.

The states completed the CAR between May and June 2000 and forwarded their data to FEMA for analysis and the subsequent development of a national report. FEMA expected the report to be completed and distributed to the President and the U.S. Congress in the first quarter of 2001.

**WHAT:** The CAR process is an initiative that is part of the EMPG continuous improvement cycle. This process is designed to assess the operations, readiness, and capabilities of a state to mitigate against, prepare for, respond to, and recover from all disasters and emergencies. The assessment focuses on the following 13 core elements that address major emergency management functions:

- Laws and Authorities
- Hazard Identification and Risk Assessment
- Hazard Mitigation
- Resource Management
- Planning
- Direction, Control, and Coordination
- Communications and Warning
- Operations and Procedures
- Logistics and Facilities
- Training
- Exercises, Evaluations, and Corrective Actions
- Crisis Communications, Public Education, and Information
- Finance and Administration

The CAR provides a common format for a self-assessment for state emergency management organizations to communicate strengths and areas needing improvement. The CAR process seeks to answer three basic questions:

- Is the emergency management program comprehensive for the needs of the states?
- Are the goals, objectives, and mission of the organization being achieved?
- Is the state able to direct strategic deployment of resources and help communities and citizens avoid becoming disaster victims?

**HOW:** Each state and territory will conduct a comprehensive self-assessment, in coordination with their respective FEMA region during Fiscal Year (FY) 2000. The results of the assessment will be used to assist states and FEMA in joint strategic planning and the identification of potential objectives for federal/state partnerships to improve their emergency management, operations, and capabilities. The CAR represents the firm commitment of both NEMA and FEMA to establish a system for assessing the capability and readiness of each state and territory to better fulfill our mission of saving lives and protecting property.

FEMA is also working with NEMA and the International Association of Emergency Management (IAEM) to develop a Local CAR that will provide local emergency managers the opportunity to evaluate their emergency management programs. It is designed to complement the State CAR to ensure greater accuracy of the results. A first draft of this document has been completed and is currently under review by NEMA, IAEM, the states, and other organizations.

*continues*

An additional important initiative is our partnership with the National Congress of American Indians (NCAI) and tribal governments to develop a Tribal Capability Assessment for Readiness (Tribal CAR). The Tribal CAR self-assessment will help tribal governments to determine the strengths and weaknesses of their emergency management programs. It will use the same format and 13 Emergency Management Functions that are found in the State and Local CAR. When completed, the Tribal CAR will be distributed to all tribal governments that are interested in improving their disaster response within tribal jurisdictions.

The development of the State CAR and ongoing work with local communities and tribal governments is managed by FEMA's Preparedness, Training and Exercises Directorate, and involved all FEMA Directorates, Offices, and Administrators of FEMA, FEMA regions, states, and territories, and NEMA. Again, this initiative has the full support of both NEMA and FEMA and is a key part of the FEMA strategy to expand the emergency management culture to focus on helping communities to avoid becoming disaster victims.

*Source:* FEMA

## PREPAREDNESS PROGRAMS

Preparedness is everyone's job. Not just government agencies but all sectors of society—service providers, businesses, civic and volunteer groups, industry associations and neighborhood associations, as well as every individual citizen—should plan ahead for disaster. As such, preparedness programs are developed to target each of these audiences in order to educate, promote, and test preparedness.

One of these public education programs is the Community and Family Preparedness Program operated by FEMA that educates the general public about disaster awareness and preparedness. The core message of the Community and Family Preparedness Program is the Family Disaster Plan—four basic steps people can take to prepare for any type of disaster.

1. *Find out what types of disasters are most likely to occur in your community and how to prepare for them.* Contacting your local emergency management office or American Red Cross chapter for information and guidelines is a good way to get started.
2. *Create a Family Disaster Plan.* Hold a family meeting to talk about the steps the family members will take to be ready when disaster happens in the community.
3. *Take action.* Each family member, regardless of age, can be responsible for helping the family be prepared. Activities can include posting emergency telephone numbers, installing smoke detectors, determining escape routes, assembling disaster supply kits, and taking first aid or CPR courses.
4. *Practice and maintain the plan.* The final step emphasizes the need to practice the plan on a regular basis so family members will remember what to do when disaster strikes.

As just one of the many preparedness programs sponsored by FEMA and other public and private disaster response and emergency management organizations, the Community and Family Preparedness Program highlights the foundation of a disaster program that is applicable to a wide range of disasters. Many more programs look specifically at preparedness regarding one type of disaster and can be obtained through agencies such as FEMA, the American Red Cross, and your state and local offices of emergency management.

## AMERICAN RED CROSS HURRICANE PREPAREDNESS TIPS

Here's what you can do to prepare for such an emergency.

### Know What Hurricane WATCH and WARNING Mean

- WATCH: Hurricane conditions are *possible* in the specified area of the WATCH, usually within 36 hours.
- WARNING: Hurricane conditions are *expected* in the specified area of the WARNING, usually within 24 hours.

### Prepare a Personal Evacuation Plan

- Identify ahead of time where you could go if you are told to evacuate. Choose several places—a friend's home in another town, a motel, or a shelter.
- Keep the telephone numbers of these places handy as well as a road map of your locality. You may need to take alternative or unfamiliar routes if major roads are closed or clogged.
- Listen to NOAA Weather Radio or local radio or TV stations for evacuation instructions. If advised to evacuate, do so immediately.

### Assemble a Disaster Supplies Kit Including the Following Items:

- First-aid kit and essential medications
- Canned food and can opener
- At least three gallons of water per person
- Protective clothing, rainwear, and bedding or sleeping bags
- Battery-powered radio, flashlight, and extra batteries
- Special items for infants, elderly, or disabled family members
- Written instructions on how to turn off electricity, gas, and water if authorities advise you to do so (Remember, you'll need a professional to turn them back on.)

*continues*

## Prepare for High Winds

- Install hurricane shutters or purchase precut 1/2-inch outdoor plywood boards for each window of your home. Install anchors for the plywood and predrill holes in the plywood so that you can put it up quickly.
- Make trees more wind resistant by removing diseased and damaged limbs, then strategically removing branches so that wind can blow through.

## Know What to Do When a Hurricane WATCH Is Issued

- Listen to NOAA Weather Radio or local radio or TV stations for up-to-date storm information.
- Prepare to bring inside any lawn furniture, outdoor decorations or ornaments, trash cans, hanging plants, and anything else that can be picked up by the wind.
- Prepare to cover all windows of your home. If shutters have not been installed, use precut plywood as described above. *Note:* Tape does not prevent windows from breaking, so taping windows is not recommended.
- Fill your car's gas tank.
- Recheck manufactured home tiedowns.
- Check batteries and stock up on canned food, first-aid supplies, drinking water, and medications.

## Know What to Do When a Hurricane WARNING Is Issued

- Listen to the advice of local officials, and leave if they tell you to do so.
- Complete preparation activities.
- If you are not advised to evacuate, stay indoors, away from windows.
- Be aware that the calm "eye" is deceptive; the storm is not over. The worst part of the storm will happen once the eye passes over and the winds blow from the opposite direction. Trees, shrubs, buildings, and other objects damaged by the first winds can be broken or destroyed by the second winds.
- Be alert for tornadoes. Tornadoes can happen during a hurricane and after it passes over.
- Remain indoors, in the center of your home, in a closet or bathroom without windows.
- Stay away from floodwaters. If you come upon a flooded road, turn around and go another way. If you are caught on a flooded road and waters are rising rapidly around you, get out of the car and climb to higher ground.

## Know What to Do After a Hurricane Is Over

- Keep listening to NOAA Weather Radio or local radio or TV stations for instructions.

- If you evacuated, return home when local officials tell you it is safe to do so.
- Inspect your home for damage.

*Source:* American Red Cross, www.redcross.org

# EDUCATION AND TRAINING PROGRAMS

Since its inception in 1979, FEMA has become a leader in developing and teaching courses in emergency management. FEMA manages the Emergency Management Institute (EMI) and the National Fire Academy (NFA), which are co-located on a former college campus in Emmitsburg, Maryland. Thousands of firefighters, fire officers, and emergency managers have been trained by FEMA. Additionally, FEMA has helped establish degree programs in junior colleges, colleges, and universities across the country. Currently, FEMA is expanding its training and education capacities through distance learning programs.

## Emergency Management Institute

FEMA defines the mission of the Emergency Management Institute as: "EMI provides a nationwide training program of resident courses and nonresident courses to enhance U.S. emergency management practices." According to the EMI catalog:

> Approximately 5,500 participants attend resident courses each year while 100,000 individuals participate in non-resident programs sponsored by EMI and conducted by State emergency management agencies under cooperative agreements with FEMA. Another 150,000 individuals participate in EMI-supported exercises, and approximately 1,000 individuals participate in the Chemical Stockpile Emergency Preparedness Program (CSEPP). Additionally, hundreds of thousands of individuals use EMI distance learning programs such as the Independent Study Program and the Emergency Education NETwork (EENET) in their home communities.

The 2001–2002 EMI Catalog of Activities listed more than 60 resident courses offered at the Emmitsburg Campus and more than 100 nonresident courses in the following subject areas:

- Mitigation
- Readiness and Technology
- Professional Development
- Disaster Operations and Recovery
- Integrated Emergency Management
- Chemical Stockpile Emergency Preparedness Program (CSEPP) Training Courses (nonresident only)

EMI also offered nearly 30 Independent Study courses in the 2001–2002 period. Some of the Independent Study courses offered include the following:

- Emergency Program Manager: An Orientation to the Position
- Radiological Emergency Management

- Hazardous Materials: A Citizen's Orientation
- Managing Floodplain Development Through the National Flood Insurance Program (NFIP)
- Animals in Disaster: Module A, "Awareness and Preparedness"
- State Disaster Management
- Multi-Hazard Planning for Schools
- The Professional in Emergency Management
- Introduction to Public Assistance Process
- Role of Voluntary Agencies in Emergency Management
- The Emergency Operations Center's Role in Community Preparedness, Response and Recovery Operations

Two courses of note offered at EMI are the Integrated Emergency Management Course (IEMC) and the Disaster-Resistant Jobs Train-the-Trainer Courses. The IEMC is a weeklong course for public officials that covers all aspects of a community emergency management function. Community officials from Oklahoma City participated in the IEMC just months before the terrorist bombing in 1995 and credit the lessons they learned at IEMC with helping them respond quickly and effectively to the bombing.

## THE INTEGRATED EMERGENCY MANAGEMENT COURSE

Protecting the population is a primary responsibility of government, and fulfilling this responsibility depends on the abilities of emergency personnel to prepare for, respond to, recover from, and mitigate against disaster. It means developing and maintaining a high standard of readiness and an ability to function effectively under crisis conditions. Emergency personnel can attain readiness either through managing emergencies or through participating in exercises. Clearly, exercises are the preferred method of gaining the necessary expertise.

The IEMC, offered by the EMI of FEMA, places public officials and emergency personnel in a realistic crisis situation within a structured learning environment. The course builds the awareness and skills needed to develop and implement policies, plans, and procedures to protect life and property through applications of sound emergency management principles in all phases of emergency management.

Community participants in IEMC include elected, midlevel management, supervisory, and operational personnel from various disciplines, including fire, emergency management, planning, finance, personnel, public health, transportation and public works, and information technology.

Early in the course, an emergency scenario begins to unfold in sequence with classroom-style lectures, discussions, and small-group workshops. As the course progresses, scenario-related events of increasing complexity, threat, and pressure occur. Participants develop emergency policies, plans, and procedures to ensure an effective response. The course culminates in an emergency exercise designed to test participant knowledge, awareness, flexibility, leadership, and interpersonal skills under extreme pressure.

Participants are challenged to use the new ideas, skills, and abilities in addition to their own knowledge and experience. In this way, the IEMC allows individuals to rehearse their real-life roles in a realistic emergency situation, while identifying additional planning needs.

*Source:* FEMA, www.fema.gov

The Disaster-Resistant Jobs course was developed in cooperation with the Economic Development Administration (EDA) of the U.S. Department of Commerce and is designed to "help small and medium-sized communities protect the economy from the effects of catastrophic events." This course was developed in response to the devastating impact the 1997 floods had on the City of Grand Forks, North Dakota. The EDA and FEMA recognized that more economic development planning could be done to reduce the impacts of future disasters on local economies.

## DISASTER-RESISTANT JOBS TRAIN-THE-TRAINER COURSES

All too often, communities that have experienced major disasters lose a major portion of their economic base. Studies have shown that after a disaster, 60 percent of small and medium-size businesses fail within two years. Many never return to business once they are closed for even a few days because of floods, tornadoes, earthquakes, and hurricanes. Not only does the community suffer from the effects of the hazard, but also in the long run the loss of jobs and tax base further reduces the community's ability to return to normal.

The EDA and FEMA have developed the Disaster-Resistant Jobs course that will help small and medium-size communities protect the economy from the effects of catastrophic events. The topics of this 3½-day course are as follows:

| Course | Topic |
|---|---|
| Unit One | The Importance of Disaster-Resistant Jobs |
| Unit Two | Creating Disaster-Resistant Jobs |
| Unit Three | Recognizing the Impact |
| Unit Four | What About Mitigation? |
| Unit Five | Disaster-Resistant Economic Development Planning Process |
| Unit Six | Business Recovery |

*continues*

The purpose of the Train-the-Trainer courses is to develop a cadre of trainers who can raise awareness in their own localities. Participants must have the desire and ability to address groups, including local economic development agencies, Chamber of Commerce meetings, service club luncheons, business meetings, and other formats to address the issue of protecting the community's economic base. Participants will be provided with a tool kit of materials that can be used to tailor their presentation before groups.

*Source:* FEMA, www.fema.gov

FEMA's EMI Higher Education Project works to establish and support emergency management curriculum in junior colleges, colleges, and universities. The project has developed a prototype curriculum for Associate Degrees in Emergency Management. Currently, FEMA lists 82 Emergency Management Higher Education Programs in institutions spread across all 50 states and Puerto Rico.

## COMMUNITY EMERGENCY RESPONSE TEAM

Following a major disaster, first responders who provide fire and medical services will not be able to meet the demand for these services. Factors such as number of victims, communication failures, and road blockages will prevent people from accessing emergency services they have come to expect at a moment's notice through 911. People will have to rely on each other for help in order to meet their immediate lifesaving and life-sustaining needs.

If it can predict that emergency services will not meet immediate needs following a major disaster, especially if there is no warning, as in an earthquake, and people will spontaneously volunteer, what can government do to prepare citizens for this eventuality?

1. Present citizens with the facts about what to expect following a major disaster in terms of immediate services.
2. Give the message about their responsibility for mitigation and preparedness.
3. Train them in needed lifesaving skills with emphasis on decision-making skills, rescuer safety, and doing the greatest good for the greatest number.
4. Organize teams so that they are an extension of first-responder services offering immediate help to victims until professional services arrive.

The Community Emergency Response Team (CERT) concept was developed and implemented by the Los Angeles City Fire Department (LAFD) in 1985. The Whittier Narrows earthquake in 1987 underscored the areawide threat of a major disaster in California. Further, it confirmed the need for training civilians to meet their immediate needs. As a result, the LAFD created the Disaster Preparedness Division to train citizens and private and government employees.

The training program that the LAFD initiated makes good sense and furthers the process of citizens understanding their responsibility in preparing for disaster. It also increases their ability to safely help themselves, their family, and their neighbors. FEMA recognizes the importance of preparing citizens. The EMI and the National Fire Academy adopted and expanded the CERT materials, believing them to be applicable to all hazards.

The CERT course will benefit any citizen who takes it. This individual will be better prepared to respond to and cope with the aftermath of a disaster. Additionally, if a community wants to supplement its response capability after a disaster, civilians can be recruited and trained as neighborhood, business, and government teams that, in essence, will be auxiliary responders. These groups can provide immediate assistance to victims in their area, organize spontaneous volunteers who have not had the training, and collect disaster intelligence that will assist professional responders with prioritizing and allocating resources following a disaster. Since 1993, when this training was made available nationally by FEMA, communities in 28 states and Puerto Rico have conducted CERT training.

The CERT course is delivered in the community by a team of first responders who have the requisite knowledge and skills to instruct the sessions. It is suggested that the instructors complete a CERT Train-the-Trainer course conducted by their State Training Office for Emergency Management or the Emergency Management Institute in order to learn the training techniques that are used successfully by the LAFD.

The CERT training for community groups is usually delivered in 2½-hour sessions, one evening per week over a 7-week period. The training consists of the following:

- Disaster Preparedness
- Disaster Fire Suppression
- Disaster Medical Operations
- Light Search and Rescue
- Disaster Psychology and Team Organization
- Course Review and Disaster Simulation

*Source:* FEMA, www.fema.gov

## National Fire Academy

The mission of the National Fire Academy (NFA) is: "Through its courses and programs, the National Fire Academy works to enhance the ability of fire and emergency services and allied professionals to deal more effectively with fire and related emergencies."

Since its inception in 1975 as the delivery mechanism for fire training for the congressionally mandated U.S. Fire Administration (USFA), the NFA estimates it has trained more than 1.4 million students. The NFA delivers courses at its Emmitsburg,

Maryland, campus, which it shares with the EMI, and across the nation in cooperation with state and local fire training organizations and local colleges and universities.

## U.S. FIRE ADMINISTRATION

As an entity of FEMA, the mission of the USFA is to reduce life and economic losses caused by fire and related emergencies, through leadership, advocacy, coordination, and support. It serves the nation independently, in coordination with other federal agencies, and in partnership with fire protection and emergency service communities. With a commitment to excellence, the USFA provides public education, training, technology, and data initiatives.

### USFA Programs

USFA programs to prevent and mitigate the consequences of fire are divided into four basic areas:

- *Public Education*. Develops and delivers fire prevention and safety education programs in partnership with other federal agencies, the fire and emergency response community, the media, and safety interest groups.
- *Training*. Promotes the professional development of the fire and the emergency response community and its allied professionals. To supplement and support state and local fire service training programs, the NFA and the EMI develop and deliver educational and training courses with a national focus.
- *Technology*. Works with the public and private groups to promote and improve fire prevention and life safety through research, testing, and evaluation. Generates and distributes research and special studies on fire detection, suppression, and notification systems, and on fire and emergency responder health and safety.
- *Data*. Assists state and local entities in collecting, analyzing, and disseminating data on the occurrence, control, and consequences of all types of fires. The National Fire Data Center describes the nation's fire problem, proposes possible solutions and national priorities, monitors resulting programs, and provides information to the public and fire organizations.

*Source:* FEMA, www.fema.gov

The NFA's on-campus programs target middle- and top-level fire officers, fire service instructors, technical professionals, and representatives from allied professions. Any person with substantial involvement in fire prevention and control, emergency medical services, or fire-related emergency management activities is eligible to apply for Academy courses. The NFA also delivers courses using CD-ROMs, their simulation laboratory, and the Internet. For those interested in pursuing degrees, the Degrees at a Distance Program extends the NFA's academic outreach through a network of seven colleges and universities. Fire service personnel who cannot attend college because of work hours and locations are able to earn a degree in fire technology and management through independent study.

### CURRICULUM OFFERED AT THE NATIONAL FIRE ACADEMY

- Arson
- Emergency Medical Services
- Emergency Response to Terrorism
- Executive Development
- Fire Prevention: Management
- Fire Prevention: Public Education
- Fire Prevention: Technical
- Hazardous Materials
- Incident Management
- Management Science
- Planning & Information Management
- Training Programs

*Source:* FEMA, www.fema.gov

### Other FEMA Education and Training Resources

FEMA provides other education and training resources such as curriculum and activities for teachers to use in the schools, school safety, and fire safety materials and information on how to talk to your kids about terrorism. FEMA has built an award-winning Website for children called "FEMA for KIDS" that has such features as becoming a disaster action kid, the disaster area, the disaster connection: kids to kids, homework help, games and quizzes, and about FEMA.

## EXERCISES

Once a plan is developed and personnel trained to the plan, the next step is exercising the plan. Exercises provide an opportunity to evaluate the efficiency and effectiveness of the plan and its components and to test the systems, facilities, and personnel involved in implementing the plan. Exercises are conducted at all levels of government and in the private sector.

FEMA defines an exercise as "a controlled, scenario-driven, simulated experience designed to demonstrate and evaluate an organization's capability to execute one or more assigned or implicit operational tasks or procedures as outlined in its contingency plan." There are four types of exercises identified by FEMA: full-scale, partial-scale, functional, and tabletop. Full descriptions of these exercise types are provided.

## EMERGENCY MANAGEMENT EXERCISE TYPES

Exercises are generally categorized by their scope:

- **Full-scale.** This exercise is used to evaluate the operational capabilities of emergency management systems over an extended period. Usually, most or all of the organization's plan will be tested. The full-scale exercise is usually conducted in conditions as close to an actual event as possible. Field teams and crews will deploy and demonstrate their procedures. The full-scale exercise is designed to stress the organization's ability to accomplish their mission under realistic conditions.
- **Partial-scale.** This is an exercise with limited goals, with a portion of the organization participating; the scope generally is less than that of a full-scale exercise. It may be conducted to evaluate a limited number of objectives or it may be used to evaluate the organization's capability to execute newly developed procedures. Some teams may be deployed to actual field sites, whereas some procedures may be demonstrated under simulated conditions. Partial-scale exercises are generally shorter than full-scale exercises.
- **Functional.** This exercise allows the evaluation of various procedures that are similar to one another, such as medical treatment or communications. It is limited to activities within a specific functional category of the organization. Activities are scenario-driven, as with the full-scale exercise.
- **Tabletop.** This exercise usually involves senior staff, elected or appointed officials in an informal setting. Using a hazard-specific scenario, supporting documentation, and injected messages simulating field-derived information, the participants discuss anticipated actions while in a controlled environment. With a facilitator to keep the discussions focused, the products derived from a tabletop exercise may include emerging policy, plan revisions, and conceptualization of new procedures.

*Source:* FEMA Comprehensive Exercise Program, July 1995

FEMA manages the Comprehensive Exercise Program (CEP). The goal of the CEP is to develop, implement, and institutionalize a comprehensive, all-hazard, risk-based exercise program. Exercises conducted under the auspices of FEMA's CEP will be used to test and evaluate emergency management plans, policies, procedures, systems, and facilities developed to mitigate against, prepare for, respond to, and recover from the effects of all types of emergencies. CEP exercises include extensive involvement of state and local officials as well as representatives from other federal agencies. The CEP program provides five categories of exercises.

## COMPREHENSIVE EXERCISE PROGRAM EXERCISE CATEGORIES

**State and Local All-Hazard Exercises.** These exercises serve as the focal point for all state and local emergency management exercise activity addressing natural, technological, and manmade disasters as well as national security hazards. They are designed to test and evaluate the operational readiness and capability of emergency management systems, identify systemic deficiencies and efficiencies, and define corrective actions needed to ensure readiness and emergency operations proficiency. Emergency management functions rather than specific scenarios will be examined.

**FEMA-Sponsored FRP Exercises.** The concept of operations, policies, and procedures set forth in the FRP for providing a federal response to state and local governments under the authorities of the Stafford Act are tested and validated in these exercises. Ideally, detailed headquarters and regional plans and procedures to implement the FRP will also be tested and validated. State and local governments will be encouraged to participate so their EOPs may be similarly tested and validated. The ultimate goal of these exercises is to achieve a seamless federal, state, and local response to and recovery from disasters of all types.

**Legislatively Mandated Exercises Supported by FEMA.** These exercises focus on plans developed at the state and local level based on guidance and requirements established by the federal government. Federal involvement in the state and local planning process is required to ensure that established standards are met and maintained. This involvement will also ensure that incorporation of hazard-specific material into the jurisdiction's single EOP is accomplished in a manner consistent with the plans of federal departments and agencies responsible for incident response.

**FEMA-Supported National and International Security Exercises.** National and international security exercises are designed to improve the capability of organizations and individuals to execute emergency management responsibilities and familiarize members of the federal government with the issues that might be encountered during a major emergency, including national security emergencies requiring the invocation of emergency authorities. These exercises also provide opportunities to validate/identify for subsequent correction, national security emergency management plans, policies, procedures, and systems. Sponsorship of these exercises is usually by the DOD or the North

*continues*

Atlantic Treaty Organization (NATO). For these types of exercises, FEMA coordinates federal civil government counterpart exercise activities.
**Special/Extraordinary Event Exercises Sponsored or Supported by FEMA.** These exercises focus on events for which overall planning rests primarily at the federal level, with other government jurisdictional elements brought in as necessary. These exercises provide opportunities to evaluate system interoperability for communications, Automated Data Processing (ADP), and other electronic media. They can also provide opportunities to explore issues and requirements for managing emergencies for which there are no plans, policies, procedures, or Memorandum of Understandings (MOUs) in existence.

*Source:* FEMA, www.fema.gov

## BUSINESS CONTINUITY PLANNING AND EMERGENCY MANAGEMENT

Business continuity planning provides focus-driven preparedness for businesses. At its simplest, business continuity planning (BCP) is the act of setting up a plan to ensure the survival of an organization. Since the early concern with the restoration of computer data, the concept of continuity has evolved in response to a changing environment. Major events have demanded that BCP encompass a growing number of concerns. The severe consequences of September 11 have raised many implications about how BCP will evolve in response to the disaster. How BCP evolves will directly influence business as a whole.

The implications of BCP are (1) terrorism must be considered as a real threat to the survival of business; (2) BCP will expand to include concern for the physical safety of employees; (3) BCP may involve the decentralization of business operations; (4) BCP may have to expand its sphere of concern to include the regional impacts of a disaster (including economic) to the area where a business is located; (5) the human relationships that a business depends on for its survival should be a major concern; (6) a recovery time of zero; (7) the renewed importance of critical data backup systems; (8) the inclusion of physical security concerns; and (9) the increased importance of and pressure on business continuity planners.

The events of September 11 raised awareness of the fact that the survival of business depends on many external factors. External factors such as infrastructure and public safety authorities play a key role in whether BCP is ultimately successful. After September 11, infrastructure vital to business has even come under the control of public safety authorities. In this case, BCP is doubly dependent on public safety authorities. This awareness has led to attempts at greater communication between business and government since the attacks. In early March 2002, the newly created Office of Homeland Security unveiled its Homeland Security Advisory System. Business immediately responded with its own proposal, the Critical Emergency Operations Link, which is intended to be a direct, two-way communication link to government at all levels. Business is demanding interaction with government so that it can anticipate how to react in the event of not only terrorist attacks, but also any catastrophe that threatens its survival. The attempt at greater communication and interaction by business is a proactive effort to turn its reliance on public safety authorities into an opportunity to ensure the success of BCP.

This approach suggests that business will demand a more extensive role for emergency management in BCP. The connection between emergency management and BCP is natural because it is the authority that has the responsibility of public safety planning. By demanding that emergency management play an extensive role in BCP, businesses can interact with government to ensure their survival. Emergency management should meet this demand with an outstretched arm because it represents a great opportunity for the field. If emergency management sincerely cooperates, then business may demand that government at all levels allocate more resources to emergency management in order to ensure that it can provide effective assistance. Ultimately, with business as its advocate, emergency management may gain the influence it needs to assume a greater role in leading the local and national public safety agenda.

# CONCLUSION

Preparedness consists of three basic elements: preparing a plan, training to the plan, and exercising the plan. Preparedness planning at the community level is critical to reducing the effects of disaster events. FEMA sponsors numerous planning, training, and education activities designed to assist communities and states in developing effective preparedness plans and training personnel to implement these plans. Through its Comprehensive Exercise Program, FEMA helps local and state governments to exercise these plans. After-action evaluation of these exercises refines the plans.

Business Continuity Planning is a significant growth area for the emergency management community. The devastating impacts of September 11 have resulted in increased coordination and cooperation between business and emergency managers. It is hoped that the emergency management community will exploit this opportunity and get businesses more active in supporting the other phases of emergency management, particularly mitigation.

# CASE STUDIES

## CASE STUDY: HOW WERE BUSINESSES AFFECTED BY THE SEPTEMBER 11 ATTACKS?

Six months later, how has BCP been affected by the attacks? The severe destruction at the World Trade Center (WTC) has led to many significant implications that are redefining BCP. In order to look at these implications, this case study first lists the latest damage estimates for businesses in the WTC and the Lower Manhattan area.

- *Death Toll.* According to a February 16, 2002 *Washington Post* article "A Towering Task Lags in New York," the attacks killed more than 2,800 people (Powell and Haughney, 2002).
- *Estimated Dollar Amount of Damage.* As of February 1, 2002, Chris Hawley writes in his article "Globalization and Sept. 11 are Pushing Wall Street off Wall Street," that the attacks caused an estimated $83 billion in damage, and only about $50 billion will be covered by insurance. Taxpayers may have to cover some of the rest (Hawley, 2002).
- *Displaced Tenants of the WTC.* According to Gary Stock of the Unblinking Website, the final tally of WTC tenants has not been completed because many sources of information contained outdated tenant lists. On the day of the attacks, the number of tenants ranged from 435 to 500. By October 19, the number increased to at least 700 (Stock, 2002).
- *Estimated Job Losses.* As of February 1, 2002, analysts predicted Manhattan would lose about 125,000 jobs after the attacks. Nearly 53,000 financial services jobs were expected to move out of Lower Manhattan—the Wall Street district—and 19,000 jobs had already left the city completely (Hawley, 2002). By February 16, 2002, one in four jobs in downtown Manhattan had disappeared—a job loss total that is thousands more than analysts had predicted immediately after September 11 (Powell and Haughney, 2002).
- *Estimate Loss of Office Space.* As of March 11, 2002, according to the article "Return to Downtown," the destruction of office space caused by the attacks equaled about 12 million square feet at the WTC and damage to another 20 million square feet in the surrounding area (Wax and Diop, 2002).
- *Communication Infrastructure Damage.* On October 29, 2001, in the article "Despite Its Losses, Verizon Went Right Back to Work Restoring Communication Services," John Rendleman writes that on the day of the attacks a Verizon switching center was destroyed by the collapse of the WTC. This caused telecom service failure to 14,000 businesses and thousands of residential customers in lower Manhattan (Rendleman, 2001). According to the article "Out of the Ashes," Verizon shared its infrastructure with some 40 competitive local exchange carriers whose services were similarly affected (Gilbert, 2002). By October 29, 2001, 90 percent of the service was restored.
- *Cleanup Concerns.* As of March 11, 2002, the cleanup of the Ground Zero site was expected to be complete by the end of May. Plans to reopen the No. 1 and No. 9 subway line stops were expected to be completed later in 2002. The reopening of the first downtown retailer was completed two weeks earlier (Wax and Diop, 2002).

A significant issue that has been raised by the devastation to office space concerns the relocation of employees. Since the attacks, 55 percent of businesses displaced by September 11 have indicated that they will return (Wax and Diop, 2002). Wax and Diop add that, "Businesses that aren't returning have largely relocated to midtown, New Jersey and elsewhere" (Wax and Diop, 2002). The issue of relocation is important given the number of employees that have moved out of the affected area. In the article "Consultants Push Wall St. to Leave," Stephen Gandel writes that, "In all, 39,610 financial services jobs have been relocated from downtown in the last six months" (Wax and Diop, 2002). More than half, 24,376 of those employees, have been moved to midtown (Gandel, 2002). In the article "Seeking Safety, Downtown Firms are Scattering," Charles V. Balgi adds, "that another 144,000 jobs are in jeopardy in a second wave of departures" (Balgi, 2002).

## CASE STUDY: CHIMACUM HIGH SCHOOL EARTHQUAKE PREPAREDNESS PROGRAM

**Program Description:** This program involves high school students teaching elementary school students about earthquake preparedness. Each class designs its own project for communicating this information. School staff see the value of such peer education.

For example, the class of 1997 designed a community service project. One element of the project was to participate in the school district's earthquake preparedness committee and provide input from the students. The students also researched the needs of classroom teachers, purchased supplies, and stocked each classroom with a "teacher's kit." They also researched and prepared personal "kid kits," which are sold for $7. The kid kits are a voluntary purchase. In addition, the students prepared an earthquake preparedness course script based on information from FEMA "Earthquake Dudes" and FEMA literature, a videotape, and an earthquake simulation with sound effects, which is available on request. Each class restocks the teacher's kit. High school students have taken American Red Cross courses, so shelters could be opened in high schools if needed.

**Evaluation Information:** Formal evaluation forms are completed after every class session by the regular classroom teacher and class students. All forms are on file. There are increased signs of school and community concern and awareness as elementary students discuss what they have learned with their parents and siblings.

**Annual Budget:** The school district budgeted $800 to $1,000 to purchase supplies for the teacher's kits.

**Sources of Funding:** The Chimacum school district and Chimacum class of 1997 fundraising

**Program Type:** Teaching earthquake preparedness

**Target Population:** Chimacum elementary school students

**Setting:** Rural Western Washington Olympic Peninsula, in a community located near a newly documented, active earthquake fault line

**Project Startup Date:** 1993

From FEMA's *Partnerships in Preparedness, A Compendium of Exemplary Practices in Emergency Management*, Volume II, May 1997

## CASE STUDY: NEIGHBORS FOR DEFENSIBLE SPACE

**Program Description:** A grassroots volunteer program, Neighbors for Defensible Space, developed out of a need to reduce the risk of uncontrolled wildfire in and around the fire-dependent district of Lake Tahoe, which has prevented catastrophic wildfires for more than 90 years.

There are three basic components in such a wildfire situation: weather, topography, and fuels. Fuels are the one element Neighbors for Defensible Space can control, and the program relies on its ability to either reduce, remove, or modify fuels. The North Lake Tahoe District program has been a model in public education and cooperative efforts in this area and has been able to demonstrate that both fire protection and environmental concerns can be addressed when dealing with wildfires. Neighbors for Defensible Space is in its second year of a five-year plan of "prescribed burning," a program that returns low-intensity fire to the forest system. In addition, the community is in the process of adopting a joint long-range master plan with its Incline Village General Improvement District, which provides water, sewage, water treatment, recreational facilities, and sanitation.

The U.S. Forest Service owns more than 650 parcels of land in the community, which has obtained approximately $900,000 in congressional funds to manage the land. In 1991 the community's taxes paid to selectively harvest 750 acres of dead and dying timber at a cost of approximately $1 million. Forty-eight percent of property owners have involved their private lands in the effort (approximately 3,500 parcels).

**Evaluation Information:** Defensible Space was recognized by the National Commission on Wildfire Disasters (a congressional committee) as a model of public education and cooperative efforts that produce results in reducing wildfire risk to urban interface communities. Their publications are used by other fire and forestry agencies.

**Annual Budget:** $5,584 in 1995 from donations

**Sources of Funding:** Primarily donations and outside agencies' earmarked funds. Local taxes, congressional funds, state forest stewardship funds, community donations, and property owners provide additional monies.

**Program Type:** Wildfire mitigation for the Reno/Lake Tahoe/Carson City region

**Target Population**: 10,000 district residents

**Setting:** Within and surrounding the Reno/Lake Tahoe/Carson City, Nevada, region

**Project Startup Date:** 1986

From FEMA's *Partnerships in Preparedness, A Compendium of Exemplary Practices in Emergency Management*, Volume II, May 1997

## CASE STUDY: SPECIAL NEEDS AWARENESS PROGRAM (SNAP)

**Program Description:** After flooding occurred in areas of southeast Texas in October 1994, students in the Community Problem Solving class of Austin Middle School, Beaumont, Texas, responded to stories they had heard about people having difficulty during emergency evacuations. The students originated the idea for SNAP and established a pilot program in their community.

The goal of SNAP is to identify those persons, such as the elderly, mentally and physically challenged, or homebound, who would have difficulty in an emergency evacuation. These residents are given special SNAP signs for display only during an emergency. SNAP also notifies police, fire, and emergency management personnel that they should look for the SNAP signs to determine where assistance is needed in an evacuation.

SNAP distributes information on the program to civic organizations, churches, and government agencies in the area through letters, speakers' bureaus, and videotapes. The program has spread throughout the United States and internationally via the Internet and magazine articles.

**Evaluation Information:** Information on the program has been requested by agencies in 31 States, the Dominican Republic, and Australia. Three magazines—*Natural Hazard Observer*, *Wanted Magazine*, and *D.E.M. Digest*—have featured articles on the program. The 41 SNAP students from Beaumont Middle School who originated the program won first place in the intermediate division in the 1995 International Future Problem Solving (Community Problem Solving) Competition in Providence, Rhode Island.

**Annual Budget:** $1,200

**Sources of Funding:** Beaumont Public Schools Foundation, Inc., FAD (Falcons Against Drugs), funds raised by SNAP team members, and personal donations

**Source for Additional Information:** Mrs. Lynne Buchwald, Austin Middle School, Beaumont, Texas (409-866-8143)

**Program Type:** Emergency evacuation assistance

**Target Population:** Elderly, physically and mentally challenged, and homebound residents who would require special assistance during an emergency

**Setting:** Any residential area in any state; the SNAP program originated in Beaumont, Texas

**Project Startup Date:** 1994

From FEMA's *Partnerships in Preparedness, A Compendium of Exemplary Practices in Emergency Management*, Volume II, May 1997

## CASE STUDY: ARCADIA CHAMBER OF COMMERCE EMERGENCY PREPAREDNESS COMMITTEE FOR BUSINESS OWNERS

**Program Description:** The Arcadia Chamber of Commerce Emergency Preparedness Committee for Business Owners provides local business owners with a disaster identification packet. The informational packet contains instructions for self-assessment of damage by the owner, along with color-coded placards that correspond to the level of need (e.g., major, moderate, or minor/no damage). Immediately following a disaster, a business owner, using the guidelines provided in the packet, would determine the extent of help needed and display the appropriate color placard. Emergency service units surveying the city would be able to instantly identify areas that required immediate assistance and thus focus available resources on those areas with the greatest need. Instructions also are provided on what supplies are needed and what activities to perform after an earthquake.

**Evaluation Information:** Other cities and counties have requested information about the disaster identification packet and indicated an interest in replicating the program. Following a presentation to the Arcadia Coordinating

Committee, the PTA expressed an interest in adapting the program for use in schools.
**Annual Budget:** None. Projects are funded individually.
**Sources of Funding:** Funds come from the Chamber of Commerce and the fire department; printing companies and manufacturers have donated printing and materials.
**Program Type:** Emergency preparedness information to help businesses identify their extent of need following a disaster
**Target Population:** Arcadia business owners
**Setting:** Arcadia, California
**Project Startup Date:** 1992
From FEMA's *Partnerships in Preparedness, A Compendium of Exemplary Practices in Emergency Management*, Volume II, May 1997

### CASE STUDY: PACIFIC GROVE, A MODEL FOR SMALL CITY DISASTER PREPAREDNESS

**Program Description:** In 1990, Pacific Grove, California (60 miles from the epicenter of the 1989 Loma Prieta earthquake), decided to prepare a comprehensive earthquake and disaster plan, following a study that showed the likelihood of a complete loss of utilities, sewer systems, and telephone services, as well as an overload of cellular systems and damage to streets and highway overpasses during an earthquake. City employees were sent to earthquake preparedness training courses given at the Governor's Office of Emergency Services' California Specialized Training Institute in San Luis Obispo. A disaster coordinator was hired to update the city's disaster plan. A Volunteers in Preparedness program was formed to train neighborhood emergency response teams, which include amateur radio operators and Boy Scouts, in earthquake preparedness, disaster medicine, how and when to turn off the gas, how to rescue victims trapped under earthquake debris, and firefighting. Lacking funding, the disaster coordinator enlisted retirement homes, volunteer organizations, public utilities, and emergency service agencies to join in the state's "Duck, Cover and Hold" earthquake drill.
**Evaluation Information:** In 1994 Pacific Grove was cited as the only city (of 12) in Monterey County having an emergency planner and the only city to hold earthquake drills regularly. Pacific Grove received the Institute of Local Self Government's California Cities Helen Putnam Award for Excellence (honorable mention, public safety) in 1995. The city's preparedness programs have received innumerable media mentions.
**Annual Budget:** $28,000 (FEMA: $11,000 toward the disaster coordinator's salary; $14,000 from the city's fire department budget; and $3,000 from the city budget)
**Sources of Funding:** FEMA and city budgets
**Program Type:** Disaster preparedness
**Target Population:** Residents of Pacific Grove (17,000)
**Setting:** Pacific Grove, California
**Project Startup Date:** 1990

From FEMA's *Partnerships in Preparedness, A Compendium of Exemplary Practices in Emergency Management*, Volume II, May 1997

## CASE STUDY: DELAWARE CITY, COMMUNITY AWARENESS AND EMERGENCY RESPONSE COMMITTEE (DC-CAER)

**Program Description:** The DC-CAER, comprising representatives of the chemical industry; volunteer organizations; and public, state, and local governments, addresses mutual concerns involving a chemical plant complex near Delaware City. Formed voluntarily in 1985, the DC-CAER strives to meet three goals: to enhance emergency response capabilities, to test and evaluate these capabilities, and to foster knowledge about chemical-related hazards and protective measures. The DC-CAER maintains a comprehensive emergency response plan to deal with chemical emergencies at the plant; conducts training programs for emergency responders; coordinates annual field emergency response exercises and tabletop drills; conducts community outreach programs to disseminate emergency information; makes presentations about its programs to community, government, and professional organizations throughout Delaware and in other states; and has produced a video that is distributed to Delaware's Extremely Hazardous Substance facilities.

**Evaluation Information:** The county has received awards from the Chemical Manufacturers Association, National Coordinating Council on Emergency Management, and U.S. Environmental Protection Agency. There have been actual emergencies without injuries.

**Annual Budget:** None, but special projects have received more than $12,000 since 1985.

**Sources of Funding:** Shared among 11 chemical plants

**Program Type:** Chemical emergency preparedness planning

**Target Population:** 6,000 residents, emergency responders, and employees and visitors of 11 chemical plants

**Setting:** Suburban environment with one small town

**Project Startup Date:** 1985

From FEMA's *Partnerships in Preparedness, A Compendium of Exemplary Practices in Emergency Management*, Volume II, May 1997

## CASE STUDY: ARLINGTON COUNTY EMERGENCY MANAGEMENT SYSTEM

**Program Description:** Arlington County's (VA) Emergency Management System was designed to provide the ability to respond to natural and/or technological disasters in a rapid and efficient manner. The system has three basic components: the Emergency Management Team (EMT), the Emergency Planning Team (EPT), and six functional task group teams. The EMT is composed of the directors of police, fire, public works, public affairs, and the County Manager's office. It is the core of the system and the decision-making body. The EPT is the think tank that anticipates future issues and makes recommendations to the EMT. The EPT and task groups brief the EMT hourly in the early stages of an incident (less frequently as the incident diminishes). During normal business, the EPT reviews the emergency operations plan to ensure that it is current. The EPT includes personnel from departments throughout the county, such as the police, sheriff, fire department, public works, public affairs, County Manager's office, parks and recreation, schools, technology and information services, and Department of Human Services. The six functional task group teams each have a different area of responsibility: shelters, communications, resources, routing and traffic control, employee support, and recovery. Members also include personnel from outside county government who have special expertise. Any of the EMT members can convene the entire team. Through the chain of command, fire and police chiefs would invoke the system. The emergency communications center would call system members, who would assemble in the emergency operations center (EOC). Each team is in a separate area of the EOC. They can communicate in person or by 800MHz radio. As an incident unfolds, the task groups monitor it on primary radio channels to anticipate resource needs and so on.

**Evaluation Information:** The program has undergone independent evaluation and has received feedback from participants in the program. Two Air Force Reserve officers, both Individual Mobilization Augmentees, have reviewed the program and participated in annual disaster exercises in which the program is evaluated. Both commented that Arlington's emergency management system was extraordinarily well developed and considerably ahead of most jurisdictions in emergency management. After each exercise, participants fill out a critique to assess their knowledge of the exercise. Results indicate a high knowledge/comfort range.

**Annual Budget:** No funds were specifically allocated for this program. The Staff Assistant to the Fire Chief was responsible for maintaining the program, so that the only outlay was a portion of his annual salary. Currently, there are only ancillary costs: printing of manuals and documents and a portion of personnel expenses.

**Sources of Funding:** Arlington County Fire Department budget

**Program Type:** Disaster preparedness and emergency management

**Target Population:** All workers and residents of the county

**Setting:** Countywide

**Project Startup Date:** 1992

# 7. The Disciplines of Emergency Management: Communications

## INTRODUCTION

Communications has become an increasingly critical function in emergency management. The dissemination of timely and accurate information to the general public, elected and community officials, and the media plays a major role in the effective management of disaster response and recovery activities. Communicating preparedness, prevention, and mitigation information promotes actions that reduce the risk of future disasters. Communicating policies, goals, and priorities to staff, partners, and participants enhances support and promotes a more efficient disaster management operation. In communicating with the public, establishing a partnership with the media is key to implementing a successful strategy.

This chapter defines the mission of an effective disaster communications strategy, outlines four critical assumptions that serve as the foundation for such a strategy, and identifies the various audiences or customers for disaster communications. The requirements for establishing a disaster communications infrastructure are defined, the difficulties in communicating risk are explored, and a strategy for communicating disaster mitigation and preparedness messages is discussed. Essential to any communications strategy is a practical guide to working with the media, which is also provided. Throughout the chapter, FEMA and the FEMA Public Affairs experiences are used as the principal example. In defining the elements of a crisis communications infrastructure used during the disaster response and recovery, the public affairs operations of FEMA are used as a model.

## MISSION

The mission of an effective disaster communications strategy is to provide timely and accurate information to the public in all four phases of emergency management:

- *Mitigation*—to promote implementation of strategies, technologies, and actions that will reduce the loss of lives and property in future disasters
- *Preparedness*—to communicate preparedness messages that encourage and educate the public in anticipation of disaster events
- *Response*—to provide to the pubic notification, warning, evacuation, and situation reports on an ongoing disaster

- *Recovery*—to provide individuals and communities affected by a disaster with information on how to register for and receive disaster relief.

# ASSUMPTIONS

The foundation of an effective disaster communications strategy is built on the following four critical assumptions:

1. Customer Focus
2. Leadership Commitment
3. Inclusion of Communications in Planning and Operations
4. Media Partnership

## Customer Focus

An essential element of any effective emergency management system is a focus on customers and customer service. This philosophy should guide communications with the public and with all partners in emergency management. A customer service approach includes placing the needs and interests of individuals and communities first, being responsive and informative, and managing expectations. The FEMA emergency information field guide illustrates the agency's focus on customer service and its strategy of getting messages out to the public as directly as possible. The introduction to the guide states the following:

> As members of the Emergency Information and Media Affairs team, you are part of the frontline for the agency in times of disaster. We count on you to be ready and able to respond and perform effectively on short notice. Disaster victims need to know their government is working. They need to know where and how to get help. They need to know what to expect and what not to expect. Getting these messages out quickly is your responsibility as members of the Emergency Information and Media Affairs team. (FEMA, 1998)

The guide's Mission Statement reinforces this point further:

> To contribute to the well-being of the community following a disaster by ensuring the dissemination of information that:
>
> - Is timely, accurate, consistent, and easy to understand
> - Explains what people can expect from their government
> - Demonstrates clearly that FEMA and other federal, state, local and voluntary agencies are working together to provide the services needed to rebuild communities and restore lives (FEMA, 1998)

The customers for emergency management are diverse. They include internal customers, such as staff, other federal agencies, states, and other disaster partners. External customers include the general public, elected officials at all levels of government, community and business leaders, and the media. Each of these customers has special needs, and a good communications strategy considers and reflects their requirements.

## Leadership Commitment

Good communications starts with a commitment by the leadership of the emergency management organization to sharing and disseminating information both internally and externally. The director of any emergency management organization must openly endorse and promote open lines of communications among the organization's staff, partners, and publics in order to effectively communicate. This leader must model this behavior in order to clearly illustrate that communications is a valued function of the organization.

In the 1990s, FEMA Director James Lee Witt embodied FEMA's commitment to communicating with the FEMA staff and partners, the public, and the media. Director Witt was a strong advocate for keeping FEMA staff informed of agency plans, priorities, and operations. Director Witt characterized a proactive approach in communicating with FEMA's constituents. His accessibility to the media was a significant departure from previous FEMA leadership. Director Witt exhibited his commitment to effective communications in many ways:

- He held weekly staff meetings with FEMA's senior managers and required that his senior managers hold regular staff meetings with their employees.
- He published an internal newsletter to employees entitled "Director's Weekly Update" that was distributed to all FEMA employees in hard copy and on the agency electronic bulletin board that updated employees on agency activities.
- He made himself and his senior staff available to the media on a regular basis, especially during a disaster response, to answer questions and to provide information.
- During a disaster response, he held media briefings daily and sometimes two to three times a day. He would hold special meetings with victims and their families.
- He led the daily briefings among FEMA partners during a disaster response.
- He devoted considerable time to communicating with members of Congress, governors, mayors, and other elected officials during both disaster and nondisaster times.
- He met four to five times per year with the State Emergency Management Directors, FEMA's principal emergency management partners.
- He gave speeches all over this country and around the world to promote better understanding of emergency management and disaster mitigation.

Through his leadership and commitment to communications, FEMA became an agency with a positive image and reputation. Communications led to increased success in molding public opinion and garnering support for the agency's initiatives in disaster mitigation.

## Inclusion of Communications in Planning and Operations

The most important part of leadership's commitment to communications is inclusion of communications in all planning and operations. This means that a communications specialist is included in the senior management team of the emergency management organization. It means that communications issues are considered in the decision-making processes and that a communications element is included in all organizational activities, plans, and operations.

In the past, communicating with external audiences, or customers, and in many cases internal customers, was not valued or considered critical to a successful emergency management operation. Technology has changed that equation. In today's world of 24-hour television and radio news and the Internet, the demand for information is never-ending, especially in an emergency response situation. Emergency managers must be able to communicate critical information in a timely manner to their staff, partners, the public, and the media.

To do so, the information needs of the various customers and how best to communicate with these customers must be considered at the same time that planning and operational decisions are being made. For example, a decision process on how to remove debris from a disaster area must include discussion of how to communicate information on the debris removal operation to community officials, the public, and the media.

During the many major disasters that occurred in the 1990s, FEMA Director Witt assembled a small group of his senior managers who traveled with him to the sites of disasters and worked closely with him in managing FEMA's efforts. This group always included FEMA's Director of Public Affairs. Similarly, when planning FEMA's preparedness and mitigation initiatives, Director Witt always included staff from Public Affairs in the planning and implementation phases. Every FEMA policy, initiative, or operation undertaken during this time included consideration of the information needs of the identified customers and a communications strategy to address these needs.

## Media Partnership

The media play a primary role in communicating with the public. No government emergency management organization could ever hope to develop a communications network comparable to those networks already established and maintained by television, radio, and newspaper outlets across the country. To effectively provide timely disaster information to the public, emergency managers must establish a partnership with their local media outlets.

The goal of a media partnership is to provide accurate and timely information to the public in both disaster and nondisaster situations. The partnership requires a commitment by both the emergency manager and the media to work together, and it requires a level of trust between both parties.

Traditionally, the relationship between emergency managers and the media has been tenuous. There has often been a conflict between the need of the emergency manager to respond quickly and the need of the media to obtain information on the response so it can report it just as quickly. This conflict sometimes resulted in inaccurate reporting and tension between the emergency manager and the media. The loser in this conflict is always the public, which relies on the media for its information.

It is important for emergency managers to understand the needs of the media and the value they bring to facilitating response operations. An effective media partnership provides the emergency manager with a communications network to reach the public with vital information. Such a partnership provides the media with access to the disaster site, access to emergency managers and their staff, and access to critical information for the public that informs and ensures the accuracy of their reporting.

An effective media partnership helps define the roles of the emergency management organizations, to manage public expectations and to boost the morale of the relief workers and the disaster victims. All of these factors can speed the recovery of a community from a disaster event and promote preparedness and mitigation efforts designed to reduce the loss of life and property from the next disaster event.

## AUDIENCES/CUSTOMERS

In order to effectively communicate disaster information, emergency managers must clearly identify their various audiences and customers. Included in many of these audiences are both partners and stakeholders. Basic emergency management audiences include the following:

- *General public*. The largest audience of which there are many subgroups, such as the elderly, the disabled, minority, low income, youth, and so on, and all are potential customers
- *Disaster victims*. Those individuals affected by a specific disaster event
- *Business community*. Often ignored by emergency managers but critical to disaster recovery, preparedness, and mitigation activities
- *Media*. An audience and a partner critical to effectively communicating with the public
- *Elected officials*. Governors, mayors, county executives, state legislators, and members of Congress
- *Community officials*. City/county managers, public works, department heads
- *First responders*. Police, fire, and emergency medical services
- *Volunteer groups*. American Red Cross, Salvation Army, the NVOADs, and so on who are critical to first response to an event

Communications with some of these customers such as the first responders is accomplished principally through radio and phone communications, as described in Chapter 4. Communicating with most of these other audiences is accomplished through briefings, meetings, provision of background materials, and, in some instances, one-on-one interviews. Communications strategies, plans, and operations should be developed to meet the information needs of each of these customers and staffed and funded accordingly.

## CRISIS COMMUNICATIONS: RESPONSE AND RECOVERY

Communicating with the public in the midst of a disaster response and recovery effort can be difficult. There are often conflicting reports on casualties and damages and usually some level of confusion among responders. Add to this situation the expectation of the public to get information almost instantaneously and the demands made by the new 24-hour news culture.

The provision of timely and accurate information directly to the public and to the media is critical to the success of any response and recovery effort. An effective communications strategy allows emergency managers and community officials at all levels of government to provide information and comfort to disaster victims and, at the same time, to manage expectations. Regular communications with the public and the media helps ensure that accurate information is being disseminated and reduces the chances for misinformation and rumors. Monitoring direct communications with victims and media reports helps identify potential problems with misinformation and rumors and allows emergency officials to address these issues before they become too widespread and damaging.

In the 1990s, FEMA built a communications infrastructure designed to disseminate critical information to the public and the media and to monitor and correct misinformation during FEMA's disaster response and recovery operations. The two key elements of FEMA's crisis communications infrastructure are staff support and technology.

## Staff Support

FEMA's Office of Public Affairs (which for a time was called the Office of Emergency Information and Media Affairs) was responsible for managing day-to-day communications activities for the agency and, during a disaster, for managing a cadre of public affairs Disaster Assistance Employees (DAEs). Public Affairs staff was responsible for establishing and managing Joint Information Centers both at FEMA headquarters and in the field and for working cooperatively with FEMA's Community Relations staff.

### Public Affairs Officers

The individuals primarily responsible for carrying out this mission are the FEMA public affairs officers (PAOs). PAOs develop and implement strategies to instill confidence in the community that all levels of government are working in partnership to restore essential services and help individuals begin to put their lives back together. They manage expectations so that disaster victims have a clear understanding of all disaster response, recovery, and mitigation services available to them. An overarching goal is to provide authoritative information to the public to combat misinformation.

### Joint Information Center

The structure FEMA uses to implement public affairs activities after a disaster is the Joint Information Center (JIC). FEMA determines the need for a JIC, and if one is established it becomes the central point for coordination of emergency public information, public affairs activities, and media access to information about the latest developments. The JIC is a physical location where PAOs from involved agencies come together to coordinate the release of accurate and consistent information to the media and the public. For a major disaster, a JIC may be established at both FEMA headquarters and on the disaster site. The on-site JIC is preferably co-located with the disaster field office. The chief spokesperson for headquarters JIC is the FEMA director of public affairs, and the chief spokesperson at the on-site JIC is the lead FEMA PAO.

### Community Relations

A partner in FEMA's public affairs operation is the Community Relations staff. The community relations function is typically performed jointly by federal and state personnel, but may also include locally hired people who know the community well. Field officers are organized into teams and deployed into affected communities to gather and disseminate information about the response and recovery operation that becomes part of the communications process. They work closely with affected states to identify community leaders and neighborhood advocacy groups to assist in disseminating information and identifying unmet needs.

## Technology

A valuable means of communications in postdisaster scenarios is the toll-free number, which has become a core element of FEMA recovery initiatives. The toll-free number is used to inform victims about the type of assistance they may be available to receive and allows them to apply for such assistance. The toll-free number is included in all forms of information and communication generated by the disaster event. An example of its usage is that during the first month after the terrorist events of September 11, 2001, more than 20,000 people called the toll-free FEMA number.

The Internet has become an increasingly popular and effective method of disseminating information to the public, and this trend will continue. FEMA's Website traffic has grown from an average of 20,000 people per week to more than 3 million. This includes users from more than 50 countries. During major disasters, the Office of Public Affairs immediately posts a special section and keeps it updated. Real-time situation reports, maps, graphics, and links to other Internet sites are posted. In addition, nearly 6,000 clients receive FEMA updates via e-mail. The interactive nature of the Internet has not yet been completely harnessed by the emergency management community and provides an opportunity to expand relationships with the public in the future.

During the 1990s, the FEMA Office of Public Affairs developed several innovative ways of disseminating information to the public. These methods have now been used in more than 200 disasters, including the Midwest floods, the Northridge earthquake, the Oklahoma City bombing, and record hurricane seasons. FEMA credits the new methods with improving its ability to get vital information out to the public and helping rebuild the agency's credibility and the nation's comfort level with its emergency management system. Some of the information dissemination methods are described as follows:

- *The Recovery Channel* provides television coverage of briefings and interviews with experts in multiple languages. Using portable satellite dishes, the signal is beamed into shelters. Network and local television news use this material. Cable television has cooperated, and a network of cable systems is committed to live Recovery Channel coverage. After the Northridge earthquake, Recovery Channel programming reached 680,000 victims on 125 cable systems in Los Angeles, with an additional potential audience of 4 million.

- *The Recovery Times* combines the latest desktop publishing technology with electronic transmission of stories and images to one printing contractor for all disasters. Prepackaging information has enabled quick publication and distribution of emergency information in an extraordinary community outreach effort. During the Midwest floods, FEMA published and distributed *Recovery Times* newspapers in nine states.
- *FEMAFAX/Spectrafax* uses the latest computerized facsimile system. Technology, comprehensive databases, and 48 telephone lines allow rapid, targeted information distribution. The system also has a fax-on-demand service. Clients select from more than 2,000 documents and material is automatically transmitted.
- *The FEMA Radio Network (FRN)* is a digitized audio production and distribution system. Radio stations can record soundbites and public service announcements with disaster officials and scientific experts. The state-of-the-art studio supports news conferences and interviews. Stations reach all of this through a toll-free number.
- *The Recovery Radio Network* distributes live broadcasts of emergency public information. It uses the Emergency Alert System (EAS) network to provide a pool feed to local radio stations that are still operating.
- *The FEMA Automatic Internet Emergency News and Situation Report Distribution Service* sends subscribers news releases and disaster situation reports via e-mail.

More information on these programs can be obtained on the FEMA Website: www.fema.gov/about/eima.htm.

## COMMUNICATING PREPAREDNESS AND MITIGATION MESSAGES

The objective of communicating preparedness and mitigation messages to the public is to educate, inform, raise awareness, and promote support for taking action before a disaster strikes.

Risk communication and public awareness programs can be undertaken in the wake of disasters or during times of normalcy. Communication of risk is an area of growing interest in the field and is discussed in more length later in this section. Public awareness is needed to gain approval for any type of emergency management measure. To implement programs, the public has to agree that a hazard exists, that it should be reduced, and that the proposed program is an appropriate measure. To achieve this consensus, the public must be involved as a partner in the process. In today's political climate, new programs are usually negotiated with the public, not decreed from officials. The case study on FEMA's Project Impact illustrates this type of approach well.

### CASE STUDY: PROJECT IMPACT

FEMA's promotion of Project Impact provides an excellent example of how to sell disaster mitigation programs to the public. The FEMA public affairs team engaged and involved the public and explained the program in terms they could understand and value, partnered with the media to get its message out, and made effective use of policy windows.

Project Impact is a community-based mitigation initiative, facilitated and partially funded by FEMA. It includes getting local businesses to partner with the local government and community organizations to prepare for and reduce the effects of future disasters. Preliminary surveys had indicated that communities were interested in reducing risk, so Project Impact was born.

The communications team's first challenge was to frame the program in terms that the public could understand. Although the program is a mitigation initiative, the team wanted to move away from emergency management jargon and describe the program in a manner with which the public would be more familiar. The slogan "put FEMA out of business" was developed. The term *mitigation* was replaced with *disaster-resistant*, and then *prevention*, and finally *risk reduction*. The slogans "prevention pays" and "prevention power" were used to reinforce the message.

A public affairs campaign was launched, both at the grassroots level within target communities and through the print and television media when possible. The communications model employed was based on the following guidelines:

- *Keep the message simple and understandable*. Literature was developed at the fourth-grade level. A "three little pigs analogy" was used to help explain the difference between preparedness and prevention.
- *Stick to the message or point*. Spokespeople used a "remember three things" tactic, whereby three main points are repeatedly mentioned in straight, clear language. Also, the Project Impact pamphlet was reduced to one page, containing five simple prevention tips.
- *Explain what's in it for the public*. The selling point to the public was that Project Impact would result in fewer losses from future disasters.
- *Educate the media on mitigation*. A media partner guide was developed to help Project Impact proponents explain to the media why mitigation is a story, why it's important, and how the media could help spread the message.
- *Involve partners*. The Salvation Army and Red Cross were solicited as partners in promoting Project Impact.
- *You are the message*. Project Impact hats and T-shirts were provided to team members.

From a media standpoint, articles were placed in the *USA Today* Op/Ed section and *Parade* magazine, and Al Roker of the *Today Show* did a spot on Project Impact. The team also took advantage of policy windows by including prevention messages into interviews during major disaster operations. Spokespeople such as FEMA's Kim Fuller promoted Project Impact in interviews during Hurricanes Irene and Floyd. An animated video on mitigation steps was provided to the networks and displayed during the interviews. Also, preprepared press releases on how people could rebuild better for the future were provided to the media.

*Source:* Interview with Kim Fuller, October 2001

## Communicating Risk

Most emergency management professionals believe that a more concerted effort to define and communicate risk to the public needs to be made. The value of warning and evacuation systems has been proven time and again but they are still often underused. Knowledge of risk does not help if the public is not informed of the danger and the actions they can take to reduce it. Bridging this knowledge gap between the scientific community and the public at large is a major area of emergency management study today.

## Risk Communication Theory

The book *Disasters by Design* by Dennis Mileti provides some valuable information on risk communication. Mileti breaks information sources for hazard awareness programs into three categories: authorities, news media, and peers. Obviously, official sources provide the most credibility. Research has shown that hazard awareness campaigns are most effective when they rely on a mix of techniques and information sources. Typically, radio and television are best for initiating or maintaining awareness, while printed materials may be best at providing detailed information.

### SEVERE WEATHER WATCHES AND WARNINGS DEFINITIONS

**Flood Watch:** High flow or overflow of water from a river is possible in the given time period. It can also apply to heavy runoff or drainage of water into low-lying areas. These watches are generally issued for flooding that is expected to occur at least 6 hours after heavy rains have ended.

**Flood Warning:** Flooding conditions are actually occurring or are imminent in the warning area.

**Flash Flood Watch:** Flash flooding is possible in or close to the watch area. Flash Flood Watches are generally issued for flooding that is expected to occur within 6 hours after heavy rains have ended.

**Flash Flood Warning:** Flash flooding is actually occurring or imminent in the warning area. It can be issued as a result of torrential rains, a dam failure, or an ice jam.

**Tornado Watch:** Conditions are conducive to the development of tornadoes in and close to the watch area.
**Tornado Warning:** A tornado has actually been sighted by spotters or indicated on radar and is occurring or imminent in the warning area.
**Severe Thunderstorm Watch:** Conditions are conducive to the development of severe thunderstorms in and close to the watch area.
**Severe Thunderstorm Warning:** A severe thunderstorm has actually been observed by spotters or indicated on radar and is occurring or imminent in the warning area.
**Tropical Storm Watch:** Tropical storm conditions with sustained winds from 39 to 73 mph are possible in the watch area within the next 36 hours.
**Tropical Storm Warning:** Tropical storm conditions are expected in the warning area within the next 24 hours.
**Hurricane Watch:** Hurricane conditions (sustained winds greater than 73 mph) are possible in the watch area within 36 hours.
**Hurricane Warning:** Hurricane conditions are expected in the warning area in 24 hours or less.

*Source:* FEMA, www.fema.gov

Different message characteristics include the amount of material, speed of presentation, number of arguments, repetition, style, clarity, ordering, forcefulness, specificity, consistency, accuracy, and extremity of position advocated. Information characteristics should be tailored for the communications goal (i.e., awareness or adoption) and for the target audience. For example, the Red Cross publishes awareness guides and manuals specific to targeted groups, such as schools, hospitals, corporations, city managers, emergency managers, and the media.

Message types vary as well. Some programs focus on content, such as scientific data or technical information about a hazard, but such information is generally processed and obtained by a small number of people. Conversely, practical instructions focus on the protective response, not the hazard itself. The simplest form of practical instruction is the "prompt," a sign that defines a single contingency and action, such as "pull lever in case of fire." Prompts are more likely to attract attention, be readily comprehended, and retained for future use. Other message styles, such as "attribute portrayal strategy," emphasize the advantages of a proposed hazard adjustment, and "fear appeals" describe the potential negative consequences of not taking the desired risk-reduction action.

Risk communication theory is based on the assumption that people leave themselves vulnerable because they are uninformed or unconvinced about the consequences of their actions. Providing accurate, helpful information would then change people's beliefs about a hazard and lead to an adoption of appropriate mitigation strategies. This is a bit of an oversimplification because many other factors and obstacles are involved, but it illustrates the general principle. The major obstacles to communicating risk and changing people's behavior include competing demands for attention, complacency, denial, and conflicts with existing beliefs.

Mileti breaks the risk communication and new behavior adoption process into the following eight steps:

1. Hearing the warning
2. Believing that it is credible
3. Confirming that the threat exists
4. Personalizing the warning and confirming that others are heeding it
5. Determining whether protective action is needed
6. Determining whether protective action is feasible
7. Determining what protective action to take
8. Taking the protective action

The field is still evolving to determine how best to influence people at each stage of the process. Most public awareness campaigns have been designed to improve disaster preparedness for near-term, high-probability threats. Less is known about what it takes to motivate people to prepare for longer-term, lower-probability events during times of normalcy. This will be an important area of study in the future.

## Risk Communication Concerns

One risk communications dilemma is how to get accurate risk information to the public when there are so many other competing, and possibly conflicting, information sources. The government has no control over what unofficial sources say because it can't regulate talking heads, so-called experts, and Websites. Partnering with the media to provide a steady stream of consistent and accurate information from responsible authorities is the best way to overcome this obstacle.

Other major issues affecting risk communication programs are when to warn the public and how much information to provide. The hurricane scenario provides the ideal model: forecasters identify the storm, watches and warnings are issued, timeframes and probabilities are provided, and the public is given clear instruction on when and how to take protective action. Communicating the risk of other hazards is not always so clear-cut, however. In the wake of the September 11, 2001, terrorist attacks, several general and unspecified terrorism threats were issued by the federal government. Weighty issues to be considered by public officials were (1) with hundreds of tips pouring in, at what point is the risk considered legitimate enough to pass on to the public, and (2) how much information on the threats should be shared.

With the first issue, officials must balance the duty to warn citizens of impending danger with concerns about unnecessarily panicking people and disrupting society. There are political and economic concerns as well. Too many false warnings could lead to a loss of credibility and public inattention to future warnings. With the second issue, officials must balance concerns about frightening the public with unthinkable rumors, and perhaps compromising important information sources, against the need to provide practical, helpful information. General, unspecified warnings may protect intelligence channels, but they do not do much to help the public prepare for the event. These are delicate issues, and a consensus on how best to responsibly educate the public about risk without unnecessarily alarming them has yet to be reached. The case study on earthquake risk in Parkfield, California, explores these issues as well.

### CASE STUDY: RISK COMMUNICATION—PARKFIELD, CALIFORNIA

One of the issues facing emergency managers is when to notify the public of a disaster risk. A desire to protect citizens must be weighed against concerns about unnecessarily alarming people, disrupting the economy, and upsetting public officials. The tension between the sheriff and the beach town mayor in the movie *Jaws* exemplifies this issue well. While the sheriff warned the mayor of the continuing risk of shark attacks, the mayor would have none of such talk during the busiest tourist weekend of the season and kept the information from the public.

A real-life scenario with a different end result, to date anyway, involved a U.S. effort at earthquake prediction in Parkfield, California, a town adjacent to the San Andreas fault. In 1985, a U.S. Geological Survey (USGS) analysis of previous earthquakes on a particular fault section indicated a strong likelihood of a repeat event by the end of the decade. The director of the USGS issued a formal public broadcast of the quake warning in April 1985 stating there was a 90 percent probability of a magnitude 5.5 to magnitude 6.0 earthquake some time between 1985 and 1993 in the Parkfield area. It also stated that a 10 percent probability existed for a magnitude 7.0 quake. By November 1988, the National Earthquake Prediction Evaluation Council (NEPEC) and the California Earthquake Prediction Evaluation Council had endorsed the prediction.

The release of the information became a national media event and precipitated a media campaign in central California involving newspapers, radio, and television that lasted years. In 1988, the California Governor's Office of Emergency Services published a detailed brochure and mailed it to 120,000 households considered at risk. It covered information about the earthquake hazard, the prediction, a possible short-term warning, and how to take action.

But the expected earthquake has not occurred. Further analysis showed that, while the successive repeat of similar but not identical quakes might be expected on individual fault sections, the amount of time between them may be highly variable. Also, confidence in predictors based on estimates of recurrence intervals has decreased in the scientific community. This case raises the issue of what to do with risk information. The duty to warn and protect the public must be balanced with fears about disrupting society with potentially unreliable risk information. It remains to be seen whether the correct decision was made for Parkfield, California.

*Source: Disasters by Design,* by Dennis Mileti

## WORKING WITH THE MEDIA

### General

The media have always been naturally drawn to disasters and emergencies because they are compelling human interest stories and provide dramatic footage. With the advent of 24-hour news stations and near real-time coverage via the Internet, the role of the media in disaster response has been magnified. In the response phase, the media often provides the most effective and efficient means for providing timely and accurate information to disaster victims and the general public. In addition, the media can play a critical role in communicating recovery information and in building support for preparedness and mitigation activities.

The biggest development in the media world over the last decade is the 24-hour news cycle. Between CNN, the major networks' all-news stations, their respective Websites, and the emergence of other independent reporting mechanisms, there is simply more air time and copy to be filled. This translates into increased coverage of disasters and emergencies and creates a demand for timely information. These pressures are only likely to grow in the future. As television becomes increasingly specialized and the number of cable channels expands, it would not be surprising within the foreseeable future to see the advent of a 24-hour *Disaster News Network*, replete with "hurricane-cams" and "on-the-fault" reporting.

The media can make a strong contribution to emergency management. Effective warnings broadcast through the media are widely credited with reducing casualties from hurricanes, tornadoes, and floods. There is often no better or quicker way to get warning messages out. The media can also facilitate assistance to disaster-stricken areas and provide reassurance to the public about the welfare of victims. Also, good science reporting can inform the public about hazards and educate them on hazard-reduction behaviors.

## Media as a Partner

Working with the media provides both a challenge and an opportunity. As discussed, the media can be a valuable element of emergency operations, disseminating important information and calling attention to urgent issues, or can be a thorn in the emergency management official's side, distributing misleading information and misguided criticism. The key to a beneficial and productive relationship is to view the media as an important partner and treat them as such.

A great example of this approach is FEMA of the 1990s. In the early 1990s, FEMA was an agency under fire, with legislators pondering its abolishment and the media producing a steady stream of criticism after a series of poorly perceived disaster response efforts. When James Lee Witt was appointed FEMA Director in 1992, he recognized communications as a key area for improvement and took appropriate measures to establish a more open and productive relationship with the media.

As a precursor to this step, the communications staff was provided with the tools and equipment needed to get the job done. Press veteran Morrie Goodman was brought on board, the office began identifying actions it could take to better partner with the media, and a host of new practices was implemented. FEMA provided the media with flyover pictures and videos from closed sites. It posted transcripts and audio of news conferences on the Internet. It created an on-site press and studio room. It provided press conferences via satellite link. It partnered with *USA Today* to include FEMA informational inserts in certain editions of the paper. The press was even provided with an area in the emergency operations center at major crises. FEMA in turn partnered with the press to promote key information to the public, such as toll-free numbers for victims to call to apply for assistance. Director Witt made it a point to constantly thank the media for their role in helping to get important messages out.

As a result of this open, collaborative approach, the public was better informed, FEMA received better press, and this translated into more support from Capitol Hill, the administration, and the public at large. Much of FEMA's success during the Clinton years can be attributed to the agency's improved ability to deal well with the press.

## MAKING INFORMATION PUBLIC AND WORKING WITH THE MEDIA

Established credibility and productive working relationships with representatives of the media is critical. In most instances, the media will be cooperative in publishing important disaster recovery information. In an ideal world, the media would simply use all news releases as issued; however, sometimes media outlets, especially in major media markets, do not use disaster recovery information as important news after the initial stories about the event. It is important to help the news media understand the important public service role they play in the recovery effort. Use the following guidelines concerning media relationships:

- Be aware of and sensitive to media deadlines.
- Respond promptly to all media inquiries. Always answer requests for information, even if only to report that the information is not available or will not be available until a given time in the future.
- Reply to questions thoroughly and accurately. Do not provide more information than is requested.
- Be honest and open. If you don't know, say so and get back to the reporter as quickly as possible with the correct answer. Ask about deadlines.
- Do not go into in-depth discussions with reporters about the programs of other agencies.
- Always be diplomatic. Especially if a request seems unreasonable, deal with it tactfully.

*Source:* FEMA Emergency Information Field Guide (condensed), October 1998

This practice extends to nongovernmental organizations (NGOs) as well. The action of the American Red Cross in the immediate aftermath of the World Trade Center terrorist attacks in 2001 provides a good example. Within a half-hour of the first plane crash, the Red Cross deployed a 35-member rapid-response team to the World Trade Center with a mission to work with media to inform the public of what was happening and what they could do to help. The Red Cross then called in a 65-member volunteer force to their offices in New York, Washington, DC, and Pennsylvania to assist with media calls. Although their persistent solicitation of aid and their subsequent plans for aid distribution eventually came under criticism, the Red Cross's immediate actions illustrate how NGOs are able to partner with the press to get important messages out.

## Managing Information

Beyond the general philosophy of treating the media as a partner, basic communications protocols must be followed. Information management is the most basic competency that must be developed. Managing information means developing a coordinated, consistent message in order to prevent confusion and maintain credibility. The release of information should be coordinated with responding partners, such as emergency management officials from other levels of government, law enforcement officials, or public health officials.

As noted earlier, FEMA achieves this through its Joint Information Center (JIC). A variation of this approach is now used by most emergency management organizations in all disaster events.

## Telling Your Own Story

Although the careful management of information flows is a critical element of any communications strategy, the desire to distribute perfect, accurate, and coordinated information must be balanced with the need to get information out quickly. The object is to tell your story before someone else tells it for you. This goal goes hand in hand with partnering with the media because the better your relationship with the media is, the more likely you will be to have this opportunity.

This was another focus of FEMA under Witt. In prior years, during major crises such as Hurricanes Iniki and Andrew, and the Loma Prieta earthquake, FEMA generally attempted to shield itself from the press while it coordinated and undertook its response and recovery activities. The resulting vacuum of information left an opening for the media to portray the FEMA response as they perceived it, and coverage of these events was generally negative toward FEMA. Conversely, during major incidents of the Witt years, such as the Midwest floods, the Northridge earthquake, and the Oklahoma City bombing, FEMA made itself as accessible to the media as possible and distributed a constant stream of information on what activities were underway and what victims could do to receive assistance. Rather than reporting on perceived deficiencies, the press shared the information with the public and FEMA's public image improved.

Another excellent example of this strategy in action is New York City Mayor Rudy Giuliani after the World Trade Center attacks. Giuliani is generally perceived as the hero of the tragedy, largely because of his effective communications via the media. He made himself constantly accessible to the press, provided continual updates on the status of response and recovery efforts, and reassured citizens that the city would rebound. By putting himself in front of the camera and articulating the story, he built public confidence and goodwill and was able to rally people together toward recovery. Even though he didn't always have all the answers, he was open, honest, and forthcoming, which fostered trust as well as good press.

The point is that if not provided with good information from good sources, the press will continue to look elsewhere. The information they find may not necessarily be accurate or fair, so it is critical to seize the communications agenda and get your story in front of the public.

### Message Objectives

The objectives of the message will obviously vary depending on the situation, but in general a media partnership can help educate, inform, reassure, and rally the public. The media can help to garner support and lay the groundwork for future emergency management measures. In times of normalcy, the media partnership can educate the public on disaster mitigation issues, although exposure may be difficult to obtain. Unfortunately, media interest in disasters is usually short-lived and does not last long into the recovery phase. Nevertheless, the media is one means of promoting mitigation with the public.

In times of a crisis or emergency, the media partnership can communicate situation reports regarding the nature and scope of the incident, the estimated human and economic damages, and what recovery measures are under way. This provides the public with a perspective of the incident and lets them know what to expect. Public officials can go on the airwaves to reassure citizens that the government is taking action and soothe the public psyche with recovery updates. Most important, the media partnership can mobilize the public toward action—whether the instruction is to call a toll-free number, evacuate homes, or open mail with gloves, there is no better way to rally the public than through the media.

## COMMUNICATIONS MEANS/PRODUCTS

### Media Lists and Contacts

FEMA's core media list consists of the following: newspapers, city and regional magazines, local trade and business publications, state bureaus of National Wire Services, local radio and television stations, local cable stations, public broadcasting stations, and public information officers at military bases. The specific contacts that an emergency management agency will typically deal with are metro desk/city reporters, public affairs reporters, business reporters, news assignment editors, and public service announcement directors.

### Press Releases

The press release is perhaps the most fundamental communications product. A press release can take the form of news releases, daily summaries, media advisories, feature articles, fact sheets, public service announcements, or other written materials. FEMA describes the objectives of its press releases to demonstrate that FEMA and its partners are working to provide critical disaster response, recovery, and mitigation programs and also to provide victims with accurate and timely information about the availability, details, and limits of these programs. FEMA press releases are routed through an established approval process.

The FEMA emergency information field guide offers some basic tips on preparing press releases. One point of emphasis for standard press releases is to never assume that information in previous disasters is appropriate for the current disaster—always review generic releases for accuracy, timeliness, and appropriateness for each specific disaster. Also, releases and advisories should be kept brief and to the point, in order to increase the likelihood that they will be used in their entirety. An example of a FEMA press release follows. It is notable for its brevity, as it concisely lists essential information such as the who, what, when, and how of victim assistance.

## FEMA PRESS RELEASE: FEDERAL DISASTER AID ORDERED FOR MISSISSIPPI STORMS

Washington, DC, December 7, 2001. The head of the Federal Emergency Management Agency (FEMA) announced today that federal disaster aid has been made available for Mississippi families and businesses victimized by tornadoes and other extreme weather that struck the state late last month.

FEMA Director Joe M. Allbaugh said the assistance was authorized under a major disaster declaration issued for the state by President Bush. The declaration covers damage to private property from the severe storms, tornadoes and flooding that began November 24.

Immediately after the President's action, Allbaugh designated the following 10 counties eligible for federal funding to help meet the recovery needs of affected residents and business owners: Bolivar, DeSoto, Hinds, Humphreys, Madison, Panola, Quitman, Sunflower, Tate and Washington.

The assistance, to be coordinated by FEMA, can include grants to help pay for temporary housing, minor home repairs and other serious disaster-related expenses. Low-interest loans from the U.S. Small Business Administration also will be available to cover residential and business losses not fully compensated by insurance.

Allbaugh said federal funds also will be available to the state on a cost-shared basis for approved projects that reduce future disaster risks. He indicated that additional designations may be made later if requested by the state and warranted by the results of further damage assessments.

Gracia Szczech of FEMA was named by Allbaugh to coordinate federal relief operations. Szczech said residents and business owners who sustained losses in the designated counties can begin the disaster application process by calling 1-800-621-FEMA, or 1-800-462-7585 (TTY) for the hearing and speech impaired. The toll-free telephone numbers will be available starting Saturday, December 8 from 8 a.m. to 6 p.m. seven days a week until further notice.

Updated: December 7, 2001

*Source:* FEMA, www.fema.gov

### Press Conferences

Press conferences allow information to be directly relayed to the media and the public. They provide officials with an opportunity to inform the public, reassure them, and mobilize them toward action. It is expected that in the aftermath of major

**Figure 7-1** FEMA Director James Lee Witt addresses the media's questions at the site of the Laguna Canyon mudflows that led to at least one death and caused a great deal of damage (February 26, 1998). Photo by Dave Gatley/FEMA.

**Figure 7-2** New York, New York, October 2, 2001. FEMA Community Relations worker answers questions from victims of the World Trade Center incident. Photo by Andrea Booher/FEMA News Photo.

crises and emergencies, elected or appointed officials will come out and show the flag via a press conference and help calm public fears. This is an important step toward recovery and a return to normalcy.

## Press Inquiries

In contrast to press releases and press conferences, press inquiries involve the media taking the communications initiative. For this reason, a dose of caution should be used when responding. The FEMA emergency information field guide provides the following general tips for interviews with the press:

- Listen to the entire question before responding.
- Avoid answering questions that call for speculation on your part.
- Be aware of false assumptions and erroneous conclusions.
- Avoid answering hypothetical conclusions.
- Be alert to multiple questions.

FEMA also has standard operating procedures to be used in receiving, responding to, and monitoring inquiries in the field. Key points of emphasis include the following:

- Never discuss program specifics or policy issues. Questions about FEMA policies or programs must always be referred to the Public Affairs Officer to be answered by the appropriate designated spokesperson.
- Ask the media to help FEMA help the disaster victims.
- Be sure to tell the media about the JIC—the single source of accurate, up-to-date, official information about the disaster.

## Websites

Websites related to emergency management have become ubiquitous. From a media communications perspective, Websites provide easy access to a repository of press releases, situation reports, general news, fact sheets, and general organizational and programmatic information. Diligence must be made to keep the site current, accurate, and easily navigable, or it loses its value as a resource. The FEMA policy for its Website (www.fema.gov) is to keep news items on the site for 30 days. The same coordination and information management practices used for press releases apply to information posted on the Internet.

## Situation Reports

Situation reports are used to provide basic information and statistics regarding emergency response efforts. The reports provide the press with facts that can be used in articles and stories and inform partner response agencies of the status of operations. FEMA produces a steady stream of situation reports in the aftermath of major events. This is consistent with the objectives of telling your story before the press does it for you and partnering with the press by being sensitive to their needs for hard data. Situation reports are typically posted on the Internet or distributed by e-mail.

An example of a typical situation report issued by FEMA during disaster response and recovery efforts is provided. Reliefweb (www.reliefweb.int/w/rwb.nsf) is an excellent source for situation reports on international crises and emergencies posted by the United Nations and other international organizations. The U.S. Office of Foreign Disaster Assistance (www.usaid.gov/hum_response/ofda/) also does a great job of providing situation reports on its assistance programs around the globe.

## FEMA SITUATION REPORT

From the Federal Emergency Management Agency (FEMA) "National Situation Update" for Tuesday, October 09, 2001 (www.fema.gov/emanagers/natsitup.htm):

### World Trade Center Update*

The City reported that as of yesterday, 393 bodies have been recovered from the World Trade Center (WTC). Of those, 335 have been identified. The number of injured is 8,786 (415 remain hospitalized) and 4,979 persons are registered as missing.

As of yesterday, 206,831 tons of debris had been removed from the WTC site (not including steel) to a landfill on Staten Island. The official estimate for total debris at the WTC is 1.4 million tons.

4,776 New Yorkers have registered for housing assistance. $9.8 million in housing assistance payments have been approved for disbursement.

3,426 New Yorkers have registered for Individual and Family Grants. $32,624 has been approved for disbursement to eligible registrants.

The Small Business Administration has approved $16,984,300 in low-interest loans to businesses and individuals.

$126,325,305 has been obligated as the Federal share for Public Assistance (as of October 8). (Manhattan DFO)

*As of October 7, 2001
*Source:* FEMA, www.fema.gov

## Spokespeople

Spokespeople can lend credibility to a message, but their words must be coordinated with the rest of the communications strategy in order to avoid multiple or contradictory messages. For this reason, it is often wise to select a single spokesperson to deliver information to the press. The lead local official is often the best person to assume this role because he or she will be best informed on the local response and the community's needs.

The FEMA press information guide for its Project Impact initiative provides some valuable pointers for spokespeople:

- Repeat information to reinforce key message points.
- Correct inaccuracies, otherwise they will be accepted as fact.
- Pair use of statistics with stories or case studies that bring them to life.
- Stay out of other people's business. Let other emergency agencies answer their own questions.
- Always be honest. If you don't know an answer to a question, say so and offer to find the answer or refer the reporter to someone who can.

## CASE STUDY: FEDERAL GOVERNMENT COMMUNICATIONS DURING ANTHRAX CRISIS

The anthrax outbreak in October 2001 provides some important communications lessons, both from the perspective of media relations and communicating risk to the public. It highlights the importance of providing a consistent, coordinated message through a single spokesperson and also highlights the need to balance a desire to reassure the public with the need to be accurate and credible.

There were two main challenges involved with the crisis. First, medical and public health officials had more questions than answers. Anthrax is a very rare disease in humans, and anthrax spores spread via the mail was basically an unknown commodity altogether. Second, there were multiple responding agencies from various levels of government involved and no established protocol for distributing information.

As a result, the public was given conflicting messages about the nature of the anthrax and misinformation about the true risk. Media criticism of the public response ensued, but it should be pointed out that in November 2001 a *USA Today* survey found that 77 percent of U.S. citizens were confident that the government could handle a major anthrax outbreak, and a Harris interactive poll showed the Centers for Disease Control's (CDC) approval rating at 79 percent. Apparently, the public was in a forgiving mood, or perhaps they were just still confused.

The first problem with the anthrax communications was that there was no clear spokesperson. A sole authority was needed to provide uniformity and consistency to the message and reduce fears. After the early conflicting messages, Tom Ridge was appointed the quasi-spokesperson for anthrax and terrorism threats, as part of his duties with the newly created Office of Homeland Security.

Beyond the issue of who should have been providing the message, there were questions about what information should have been provided. The case illustrates a classic communications conundrum. Officials were under pressure to provide current information to the public, which was seeking reassurances, while there was still much uncertainty about the true nature of the threat. Marc Shannon, director of Ketchum's Washington D.C. healthcare practice summed up the dilemma well: "If you don't get out enough information you're accused of being secretive. And if you give too much information you are criticized for stirring up anxiety." As Shannon points out, a key to communications in these instances is not to be afraid to say I don't know.

Tommy Thompson of the Department of Health and Human Services might be accused of erring in this respect. During an interview on *60 Minutes* early in the crisis he said, "We've got to make sure that people understand that they're safe, and that we're prepared to take care of any contingency, any consequence that develops from any kind of bio-terrorism attack." After new cases of anthrax continued to be reported, and two D.C. postal workers and a Connecticut woman later died of inhalation anthrax, it became apparent that this was a case of an official going too far in trying to assuage public fears. These remarks were in contrast to those of New York City Mayor Rudy Giuliani, who after the death of the Connecticut woman said words to the effect that the government can't guarantee that every single person will be completely safe from anthrax, and that individuals need to exercise a certain amount of due diligence. Although these remarks may not have been completely comforting, they were accurate, practical, and fostered public trust.

*Source: PR Week*

## CONCLUSION

Whether dealing with the media, the public, or partners, effective communication is a critical element of emergency management. Media relations should be open and cooperative, the information stream must be managed to provide a consistent, accurate message, and officials need to be proactive about telling their own story before it is done for them. A customer service approach is essential to communicating with the public, a collaborative approach should be taken to promoting programs, and great care should be given as to how and when risk is communicated to citizens.

# 8. International Disaster Management

## INTRODUCTION

People of all nations face risks associated with the natural and technological hazards described throughout this book, and almost all nations eventually become victim to disaster. Throughout history, civilizations have adapted to their surroundings in the hopes of increasing the likelihood of survival. As societies became more organized, complex systems of response to these hazards were developed on local, national, and regional levels. The capacity to respond achieved by individual nations can be linked to several factors, including propensity for disaster, local and regional economic resources, organization of government, and availability of technological, academic, and human resources; however, it is becoming increasingly common that the response ability of individual nations is insufficient in the face of large-scale disaster, and outside assistance must be called upon. Disasters that affect whole regions are not uncommon and require these same international response mechanisms.

This chapter introduces the conglomeration of agencies, including the U.S. government, international organizations, nongovernmental organizations (NGOs), and financial institutions, that prepare for and respond to the natural, technological, and complex humanitarian emergencies (CHEs) that overwhelm the capacity of any one sovereign nation. The mission and goals of each of these entities and groups are described (although their performance is not detailed). In conclusion, a comprehensive case study is presented on the international response to the Gujarat, India earthquake of January 26, 2001.

## DISASTERS IN DEVELOPING NATIONS

Disasters of all kinds strike literally every nation of the world, although these events do not occur with uniformity of distribution. The developing nations suffer the greatest impact of nature's fury, and these same nations are also most often subject to the internal civil conflict that leads to CHEs. Furthermore, the greatest incidence of natural disasters occurs within developing countries, with 90 percent of disaster-related injuries and deaths sustained in countries with per-capita income levels that are below $760 per year (UNICEF).

Although disaster preparedness and mitigation are widely accepted by international development agencies to be integral components in the overall development process, it comes as no surprise that countries ranking lower on development indices have placed disaster management very low in budgetary priority. These nations' resources tend to be focused on more socially demanded interests such as education and base infrastructure, or on their military, instead of on projects that serve a preparatory or mitigative need, such as retrofitting structures with hazard-resistant construction. Because disasters are chance events, and thus not guaranteed to happen, disaster management programs in poor countries tend to be viewed as superfluous. Delegating disaster management responsibilities to the military is also commonly seen even in countries with a moderate level of development, although these agencies are rarely trained to carry out the necessary response tasks required. To compound the situation further, poverty and uncontrolled urbanization often force large populations to concentrate in perilous, high-risk urban areas that contain little or no defense against disasters.

# INTERNATIONAL INVOLVEMENT

A disaster requires the involvement of the international community of responders when a nation's capability to respond has become overwhelmed. This threshold is determined by many factors, including the availability of economic resources, the level of local responder training, the resilience of the infrastructure, the public opinion of the government's ability to manage the crisis, and the availability of specialized assets, among many others. Of course, this threshold is crossed much earlier in the poorer countries. It must be recognized, however, that even the wealthiest nations regularly find themselves in need of help from the international community, whether for supplies, manpower, money, or a specific skill or asset that cannot be found locally. Appeals for assistance are made in many ways and are often simultaneously met with unsolicited offers of aid and support. With the global interconnectivity brought about through television and the Internet (the so-called CNN effect), news of a disaster can circle the globe within minutes, stirring the machine of response into action.

There are three types of emergencies that normally involve an international humanitarian response: natural disasters, technological disasters, and complex humanitarian emergencies (CHEs). While the first two are clearly defined, the CHEs have been subject to diverse interpretations and changing standards, and thus, for the purposes of this book, are characterized by the definition established by the United Nations (UN). They classify a CHE to be a "humanitarian crisis in a country or region where there is total or considerable breakdown of authority resulting from the internal and/or external conflict and which requires an international response that goes beyond the mandate or capacity of any single agency." (DODCCRP) Andrew Natsios, Director of the U.S. Agency for International Development (USAID), identifies five characteristics most commonly seen in CHEs in varying degrees of intensity:

- Civil conflict, rooted in traditional ethnic, tribal, and religious animosities (usually accompanied by widespread atrocities)
- Deteriorated authority of the national government such that public services disappear and political control dissolves

- Mass movements of population to escape conflict or search for food, resulting in refugees and internally displaced people (IDPs)
- Massive dislocation of the economic system, resulting in hyperinflation and the devaluation of the currency, major declines in gross national product, skyrocketing unemployment, and market collapse
- A general decline in food security, often leading to severe malnutrition and occasional widespread starvation (Natsios, 1997)

Although these emergencies are fundamentally different from natural and technological disasters in regards to their generally political and intentional sources, they share many characteristics in terms of their requirements for response and recovery. In accordance, many of the organizations and entities described in this chapter respond to all three types of disasters indiscriminately.

## IMPORTANT ISSUES INFLUENCING THE RESPONSE PROCESS

Several issues must be addressed when responding to international disasters. The first, *coordination*, is a vital and immediate component because of the sheer numbers of responding agencies that almost always appear. It is not uncommon in larger disasters to see several hundred local and international NGOs, each with a particular skill or service to offer. Successful coordination and cooperation can lead to great success and many lives saved, but infighting, turf battles, and nonparticipation can lead to confusion and even cause a second disaster (PAHO).

The UN has become widely recognized as the central coordinating body, with specialized UN agencies handling the more specific needs associated with particular disaster consequences. Most often, the UN capitalizes on longstanding relationships with the host country to form a partnership on which they establish joint control. In addition to the UN, several organizations and associations have come up with standards of conduct, such as the Red Cross Code of Conduct (www.ifrc.org/publicat/conduct/index.asp), the Sphere Project Humanitarian Charter and Minimum Standards in Disaster Response (www.sphereproject.org/handbook_index.htm), and the Oxfam Code of Conduct for NGOs (www.oxfam.org).

The second issue is that of *sovereignty of the state*. State sovereignty is based on the recognition of political authority characterized by territory and autonomy. Accordingly, a foreign nation or organization cannot intercede in domestic matters without the prior consent of the ruling government. This can be a major hurdle in CHEs that have resulted from civil war, such as the peacekeeping mission in Somalia where there existed no official government in place with which to work. Although not as commonly seen, sovereignty has also been an issue in matters of natural and technological disasters, particularly when a nation does not want to be viewed as weak or unable to take care of its people. Examples of such behavior include Japan's refusal to allow access to international agencies for several days after the earthquake in Kobe and the actions of the Former Soviet Union following the nuclear power plant accident in Chernobyl.

The third issue is *equality in relief distribution*, and it applies to any type of disaster. Situations often arise where, for any number of cultural or political reasons, certain groups in need of aid are favored over others. The first example of this discrimination is the result of gender bias, which is most commonly found in societies where gender roles are strictly defined and women are traditionally tasked with duties related to the home and children (which tend to be increased in times of crisis). In these cultures, the men are more likely to have opportunities to wait in relief lines for supplies, and the women (as well as children and the elderly) become even more dependent on them for survival. This situation is exacerbated if a woman is a widow or single parent and has no ability to compete for distributed aid.

The second form of inequality in relief is that of class bias. Although most obvious in social systems explicitly based on caste identity, underlying ethnic and racial divides often present similar problems. Avoiding these forms of bias is difficult because the agencies involved must be aware of the discrimination in order to counteract its influence. Often, host-country nationals are "hired" by humanitarian agencies to assist in relief distribution, and inadvertent hiring of specific ethnic or social groups can lead to unfair distribution along those same ethnic/social lines. At the same time, humanitarian agencies are quick to focus on those groups most visibly affected by a CHE, such as IDP populations, causing an inordinate percentage of aid to be directed to them, while other needy groups go unnoticed.

Many of the international response agencies are continuously developing systems of relief and distribution that work to counteract the complex problems associated with these biases; however, the difficult nature of this issue is highlighted in the fact that specifically targeting groups, such as women or children, can lead to reverse discrimination. Any of these biases can lead to a decline in perceived legitimacy or impartiality of the assisting agency and/or result in exacerbation of the needs being addressed (Maynard).

A fourth issue is the importance of *capacity building* and *linking relief with development*. Responding agencies have an obligation to avoid using a bandage approach in assisting the affected country. Disasters almost always present a window of opportunity to rebuild old, ineffective structures and develop policy and practice in a way that leaves behind a more empowered, resilient community. Because these goals mirror those of most traditional development agencies, linking relief and development should not be a major deviation from either type of agencies' missions. These opportunities are greatest in situations that require the complete restoration of infrastructure and basic social services, and are found equally in disaster and CHE scenarios. In the reconstruction phase, it is vital that training and information exchanges occur and that local risk is fully incorporated to mitigate for repeat disasters. These repeat disasters often contribute greatly to a nation's lag in development, and therefore fully addressing them is vital to increasing the nation's likelihood of being developed sustainably.

## THE UNITED NATIONS SYSTEM

The UN began in 1945, when representatives from 51 countries met in San Francisco to establish the United Nations Charter as a commitment to preserve peace in the aftermath of World War II. Later that year, the Charter was ratified by the five permanent members: China, France, the Soviet Union, the United Kingdom, and the United States, as well as several other countries. Today, 189 countries are members of the UN, and the Charter (which is similar to a sovereign state's constitution and establishes the rights and responsibilities of Member States) is amended as is necessary to reflect the changing needs of current world politics.

The UN itself is not a government body, nor does it write laws; however, the autonomous Member States do have the ability through the UN to resolve conflict and create international policy. No decision or action can be forced on a sovereign state, but as global ideals are naturally reflected through these collaborative policies, they are usually given due consideration.

Through the major UN bodies and its associated programs, the UN has established a presence in most countries throughout the world and fostered partnerships with Member State governments. Although more than 70 percent of UN work is devoted to development activities, several other issues are central in their mission, including disaster mitigation, preparedness, response, and recovery. In the event of a disaster, the UN is quite possibly the best equipped to coordinate disaster relief and to work with the governments to rehabilitate and reconstruct. This is especially true in the case of the developing countries, where regular projects are ongoing and must be adjusted to accommodate for damages to infrastructure and economy caused by recurrent disasters, and where disasters quickly exhaust the response capabilities.

Upon onset of a disaster, the UN responds immediately and on an ongoing basis by supplying aid in the form of food, shelter, medical assistance, and logistical support. The UN Emergency Relief Coordinator heads the international UN response to crises through a committee of several humanitarian bodies, including the UN Children's Fund (UNICEF), the UN Development Programme (UNDP), the World Food Programme (WFP), the UN High Commissioner for Refugees (UNHCR), and other associates as deemed necessary in accordance with the problems specific to the event. Each of these agencies, as shown in this section, fulfills a specific need presented by most humanitarian emergencies, be they natural or manmade.

The UN also promotes prevention and mitigation activities through its regular development projects. By encouraging the building of early warning systems and the conducting of monitoring and forecasting routines, they are working to increase local capacity to adequately boost local and regional preparedness. In the conclusion of the International Decade for Natural Disaster Reduction of the 1990s (which strove to focus on a shift from disaster response-oriented projects to disaster mitigation), the UN adopted its International Strategy for Disaster Reduction (ISDR) to promote the necessity of disaster reduction and risk mitigation as part of its central mission. This initiative seeks to enable global resilience to the effects of natural hazards in order to reduce human, economic, and social losses, through the following mechanisms:

- Increasing public awareness
- Obtaining commitment from public authorities
- Stimulating interdisciplinary and intersectoral partnership and expanding risk-reduction networking at all levels

- Enhancing scientific research of the causes of natural disasters and the effects of natural hazards and related technological and environmental disasters on societies

These strategies are carried out through the country offices and local governments, in the most vulnerable communities. Mitigation and preparedness strategies are implemented at all levels of society via public awareness campaigns, secured commitment from public authorities, intersectoral cooperation and communication, and technical knowledge transfer.

## The United Nations Development Programme

The United Nations Development Programme (UNDP) was established in 1965 during the UN Decade of Development to conduct investigations into private investment in developing countries, to explore the natural resources of those countries, and to train the local population in development activities such as mining and manufacturing. Over the years, as the concept and practice of development expanded greatly, the UNDP took on much greater responsibilities within host countries and within the United Nations as a whole.

Historically, the UNDP was not considered an agency on the forefront of the crisis and disaster management scene because, although they worked on development issues, they did not focus specifically on emergency response systems, which were considered to be the focal point of crisis and disaster management for many years. As mitigation and preparedness received greater emphasis in the field, however, the vital role that the UNDP has played all along is being increasingly recognized. Capacity building has always been central to the mission of the UNDP, in terms of empowering host countries to be better able to address issues of national importance, eventually without foreign assistance.

In the execution of UNDP projects, there was a natural, although unintended, move toward activities that indirectly filled mitigation and preparedness roles. Projects that worked to strengthen government institutions also improved the capacity of such institutions to respond with appropriate and effective policy, power, and leadership in the wake of a disaster. By its very nature, therefore, capacity building could clearly be considered a mitigation activity (although early on, the mitigation of disasters was not as widely understood or practiced as was the response to them).

Attention to disaster management increased through time as natural and manmade disasters were affecting greater populations and causing greater financial impacts, and the developing nations felt the greatest inability to prepare and/or respond to them. It was widely recognized that the unguided development trends typified by these lesser-developed nations led to their greater vulnerabilities. For example, developing countries generally have a severe deficiency in physical infrastructure from which response could be based, they participate in environmental misuse and destruction that exacerbates certain natural hazards, and they often contain migrant populations that settle in concentrated groups within disaster-prone regions.

Considering that 90 percent of natural disasters occur in developing countries, and likewise that 90 percent of victims of disasters live in developing countries, it becomes apparent that the issue of their management cannot be kept separate from the mission of the UNDP (which works primarily in these countries).

Today, the UNDP believes that vulnerability to disasters is strongly linked to a lack of or weak infrastructure, poor environmental policy, misuse of land, and rising populations in areas that are prone to repeat disasters. In many cases, these disasters can literally set a country back years, if not decades, in terms of development achievement. For instance, the president of Honduras has declared that the country has gone back to early 1950s levels of development because of the devastating effects of Hurricane Mitch. It is also recognized that small to medium-size disasters in the least developed countries, can "have a cumulative impact on already fragile household economies and can be as significant in total losses as the major and internationally recognized disasters" (UNDP). It is their modern objective, then, to "achieve a sustainable reduction in disaster risks and the protection of development gains, reduce the loss of life and livelihoods due to disasters, and ensure that disaster recovery serves to consolidate sustainable human development" (UNDP).

In 1995, as part of the UN's changing approach to better assisting the relief community as a whole, the Emergency Response Division (ERD) was created within the UNDP. This move drastically augmented the organization's role in responding to disasters. Additionally, 5 percent of UNDP budgeted resources were allocated for quick-response actions in special development situations by ERD teams, thus drastically reducing delays in bureaucratic decision making. Specifically, the ERD helps in creating a collaborative framework among the national government, UN agencies, donors, and NGOs that immediately respond to disasters, provides communication and travel to disaster management staff, and distributes relief supplies and equipment. Following the deployment of ERD teams (generally 30 days), a detailed project plan is submitted, and a full UNDP project can be applied to a disaster. ERDs work in strengthening coordinating mechanisms, and their central strong organization role has shaped future UNDP involvement in disasters.

In 1997, under the UN Programme for Reform, the responsibilities and operational activities of the Emergency Relief Coordinator, regarded as being part of national capability or capacity, were formally transferred to the UNDP. In response, the UNDP created the Disaster Reduction and Recovery Programme (DRRP) within the ERD. The broad-ranging duties of this program pertaining to disaster mitigation, prevention, and preparedness were defined as follows:

- Mainstream disaster reduction into development policy, strategies, plans, and programs
- Strengthen capacity of institutions at all levels for enhanced disaster management
- Develop innovative approaches to accelerate sustainable postdisaster recovery, promoting the inclusion of disaster reduction measures into rehabilitation and reconstruction
- Build partnerships, promote networks, and facilitate cooperation at international, regional, and national levels
- Facilitate the development and delivery of high-quality training and human resource development activities

- Promote and develop disaster-reduction policies and strategies
- Represent UNDP at interorganizational fora on the topic
- Provide direct substantive support to multisectoral integrated country programs

In terms of response, recovery, and reconstruction, the UNDP pledged to support activities in eight subject groups that handle emergencies of natural, technological, and CHE disasters. The groupings are such that any combination can be applied to meet the needs of virtually any type of emergency situation that arises. They are listed as follows, with their summarized roles included:

- *Emergency Interventions*. Establish the nature and scope of the emergency, collect and distribute timely information to all parties involved, and track and coordinate donations from domestic and foreign sources. As the UNDP resident representative of the country leads the effort (through the Disaster Management Team), there is long-term institutional knowledge to manage the disaster response.
- *Programming for Peace and Recovery*. Determine the major difficulties to be addressed and the priority needs in terms of external support, assist the current or new government in addressing these issues, and provide the planning and financial coordination that is required. What is most unique about the UNDP activities is that they seek to break the dependence that has been created by outpourings of international relief that can hamper a return to normalcy after natural or manmade disasters.
- *Area Rehabilitation to Resettle Uprooted Populations*. Create or expand on the capacity of the communities where IDPs or refugees are to be resettled. The UNDP utilizes many of its standard project schemes, such as creation of income-generating activities, building stronger infrastructure, and promoting local participation in the process.
- *Reintegrating Demobilized Soldiers*. Initiate reintegration projects and coordinate the funding from the international community. The UNDP has been able to provide much of the administrative duties of this task and the follow-up once the agencies more specifically concerned with demobilization have left. In the 1990s alone, the UNDP channeled more than $150 million to reintegration and demobilization programs.
- *Demining*. Provide general management input for the conduct of operations and coordinate financial contributions from international donors. Because it is necessary to clear mines before development can continue with any chance of success, the UNDP sees this task as integral in their goal of linking relief to development.
- *Rebuilding Institutions and Improving Government*. Make overall assessments of the state of governance, identify problems that need to be addressed, and assist the government in the coordination of reform/restructure/repair. The long partnerships and the assumed neutrality make the UNDP an ideal body for this role.
- *Organizing National Elections*. Provide local coordination and technical assistance. In many cases, UNDP involvement can give a sense of legitimacy to an

election at a time when stability is fragile, especially after a new government has come to power after a civil conflict. This stability is vital if the country is to emerge from its crisis.

- *Managing Delivery of Program Aid.* Manage UN Office for Project Services in the delivery of program aid, assist in the procurement of services, and assist in the administration of loans. "The donors themselves, bilateral and multilateral, are not often coordinated at the central level where major decisions on allocation of funds are usually made."

In addition to the aforementioned roles and responsibilities, the UNDP leads several interagency working groups. One such group, which consists of representatives from the World Food Programme (WFP), the World Health Organization (WHO), the Food and Agriculture Organization (FAO), the UN Populations Fund (UNFPA), and the UN Children's Fund (UNICEF), works to develop principles and guidelines in order to incorporate disaster risk into the Common Country Assessment and the UN Development Assistance Framework. They have included in their goals capacity building for the central governments consulted, assessment of vulnerability, creation of early warning systems, development and maintenance of a framework for contingency planning, greater efforts toward mine removal, strengthening of country disaster management programs and teams, and national development programs that include the all-hazard spectrum.

The ISDR Working Group on Risk, Vulnerability and Disaster Impact Assessment works on the setting of guidelines for social impact assessments. The UNDP also coordinates a Disaster Management Training Programme (DMTP) in Central America, which runs a conference on "The use of microfinance and microcredit for the poor in recovery and disaster reduction," and has created a program to elaborate financial instruments to enable the poor to manage disaster risks.

The UNDP currently dedicates more than 40 percent of its resources to emergency relief operations. It is clear that they are not a self-contained disaster relief organization, but that is not how the UN system was developed to function. The many agencies and offices that are involved would not act efficiently without a central coordinating body, and the UNDP has recently been deemed the most able to handle that duty. As should be clearly presented, these duties do not strain the established role as lead development agency within the UN system.

## The United Nations Office for the Coordination of Humanitarian Affairs

The United Nations Office for the Coordination of Humanitarian Affairs (OCHA) was created under the UN Secretary-General's Program for Reform in 1998, to accommodate the needs of victims of disasters and emergencies. Their specific role in the broad range of disaster management tasks is to coordinate assistance provided by the UN system in emergencies that exceed the capacity and mandate of any individual agency. The OCHA response to disasters can be categorized under three main groupings, including coordinating the international humanitarian response, providing support and policy development to the humanitarian community, and advocating humanitarian issues to ensure that the overall direction of relief reflects the general needs of recovery and peace-building.

The head of OCHA is the UN Emergency Relief Coordinator (ERC) and is responsible for the coordination of the response efforts of the UN through the Inter-Agency Standing Committee (IASC). The IASC is a group consisting of both UN and outside humanitarian organization leaders, which analyzes crisis scenarios and formulates a joint response to ensure maximum effectiveness and minimal overlap of relief. The ERC works to deploy appropriate personnel from throughout the UN to assist the resident coordinators and lead agencies in the response, thus increasing the likelihood that on-site coordination will be strong.

The Disaster Response System, established by OCHA, constantly monitors the onset of natural and technological disasters. This system includes training the assessment teams before disasters strike, as well as evaluations conducted postdisaster. When a disaster is identified, the OCHA response is activated, and a situation report is generated to provide the international response community with detailed disaster-specific information (which includes damage caused, actions taken, needs assessed, and current assistance being provided). OCHA may then, if deemed necessary, deploy a UN Disaster Assessment and Coordination (UNDAC) team to assist in the coordination of relief activities and help assess damages and needs (these duties are not as overreaching as in complex emergencies; see the previous UNDP section).

An Operations Coordination Center may be set up in the field in order to assist local first-response teams in their coordination of the often overwhelming international representation of relief agencies that respond. Finally, OCHA can set up communications capabilities if they have been damaged or do not exist to the capacity required by the UN responding agencies. OCHA responsibilities are generally concluded when the operation moves from response to recovery.

## The United Nations Children's Fund

Like most other major UN agencies, the UN Children's Fund (UNICEF, formerly known as the United Nations International Children's Emergency Fund) was established in the aftermath of World War II. Its original mandate was to aid the children suffering in postwar Europe, but its mission has been expanded to address the problems that affect poor children throughout the world. UNICEF is mandated by the General Assembly to serve as an advocate for children's rights, to ensure that each child receives at least the minimum requirements for survival, and to increase their opportunities for a successful future. Under the Convention on the Rights of the Child (CRC), a treaty adopted by 191 countries, the UNHCR holds wide-reaching legal authority to carry out its mission.

Before the onset of disasters, it is not uncommon for UNICEF to have established itself as a permanent in-country presence, with regular budgetary resources. In the situations of disaster or armed conflict where this is the case, UNICEF is well poised to serve an immediate role as aid provider to its specific target groups. This rapid response is important because young mothers and children are often the most marginalized groups in terms of aid received. UNICEF works on a regular basis to ensure that children have access to education, healthcare, safety, and protected child rights. In the response and recovery periods of humanitarian emergencies, these roles are merely expanded to suit the rapidly extended requirements of victims. In countries where UNICEF has not yet established a permanent presence, the form of aid is virtually the same; however, the timing and delivery are affected, and reconstruction is not nearly as comprehensive.

UNICEF maintains that humanitarian assistance should include programs aimed specifically for child victims. Relief projects generally work to provide a rapidly needed response in the form of immunizations, water and sanitation, nutrition, education, and health. Women are recipients of this aid as well because UNICEF considers them to be vital in the care of children. UNICEF also works through recovery and reconstruction projects, providing for the basic rights of children. UNICEF is currently working in 161 countries.

## The World Food Programme

The World Food Programme (WFP) is the arm of the UN tasked with reacting to hunger-related emergencies throughout the developing world. The WFP was created late in 1961 by a resolution adopted by the UN General Assembly and the UN Food and Agriculture Organization (FAO). Chance enabled the program to prove the necessity of their existence when the WFP provided relief to more than 5 million people several months before they were deemed officially operational in 1963. In the year 2000 alone, the WFP fed 83 million people through its relief programs. Over the course of its existence, the WFP has provided more than 43 million metric tons of food to countries worldwide.

Because food is a necessity for human survival, it is a vital component of development. The WFP works throughout the world to assist the poor who do not have sufficient food to survive "to break the cycle of hunger and poverty." Hunger alone can be seen as a crisis because more than 800 million people across the globe receive less than the minimum standard requirement of food for healthy survival. Hunger is often associated with other crises, including drought, famine, and human displacement, among others.

In rapid-onset events such as natural disasters, the WFP is activated as a major player in the response to the immediate nutritional needs of the victims. Food is transported to the affected location and delivered to storage and distribution centers. The distribution is carried out according to preestablished needs assessments performed by OCHA and the UNDP. The WFP distributes food through contracted NGOs who have vast experience and technical skills required to plan and implement such projects of transportation, storage, and distribution. The principal partners in their planning and implementation are the host governments (who must request the aid of the WFP to begin with, unless the situation is a CHE where there is no established government, and the UN Secretary General makes the request). The WFP works closely with all responding UN agencies to coordinate an effective and broad-reaching response because food requirements are so closely linked to every other vital need of disaster victims.

In the aftermath of disasters, during the reconstruction phase, it is often necessary for the WFP to remain an active player through continued food distribution.

Rehabilitation projects are implemented in a way that fosters increased local development and include providing food aid to families, who as a result will have extra money to use in rebuilding their lives, and food for work programs, which break the chains of reliance on aid as well as provide an incentive to rebuild communities.

### The World Health Organization

The idea for the World Health Organization (WHO) was proposed during the original meetings to establish the UN system in San Francisco in 1945. In 1946, at the United Health Conference in New York, the WHO constitution was approved, and on April 7 (World Health Day), it was signed and made official. Like the WFP, WHO proved its value by responding to an emergency (a cholera epidemic in Egypt) months before it was an officially recognized organization.

WHO was established to serve as the central authority on sanitation and health issues throughout the world. They work with national governments to develop medical capabilities and health-care and assist them in the suppression of epidemics. WHO supports research for the eradication of disease and provides expertise on these subjects when requested. They provide training and technical support and develop standards for medical care.

In the event of a disaster, WHO responds in several ways that address the health of victims. Most important, it provides ongoing monitoring of diseases traditionally observed within the unsanitary conditions of disaster aftermath. WHO also provides technical assistance to the responding agencies and host governments who are establishing disaster medical capabilities and serves as a constant source of expertise as needs arise.

Since its inception, regional offices have been established. These offices, which comprise the 191 separate member states of WHO, focus on the health issues most directly related to each regional area's needs and concerns. These regions include the following:

- African Regional Office (AFRO)
- Pan American Health Organization (PAHO)
- South-East Asia Regional Office (SEARO)
- Regional Office for Europe (EURO)
- Eastern-Mediterranean Regional Office (EMRO)

## NONGOVERNMENTAL ORGANIZATIONS

The number of nongovernmental organizations (NGOs) focusing on international humanitarian relief has grown exponentially in the past few decades. These organizations have come to play a vital role in the response and recovery to disasters, filling gaps left by national and multilateral organizations. They have significantly improved the ability of international relief efforts to address the needs of victims with a diverse range of skills and supplies. Some of the larger NGOs, like the International Committee of the Red Cross (ICRC), have established an international presence similar to that of the UN and have developed strong local institutional partnerships and a capacity to respond almost immediately with great effectiveness. These grassroots-level organizations are so successful in their activities that the major funding organizations such as USAID, OFDA, and the UN regularly arrange for relief projects to be implemented by them rather than their own staff.

There are several classifications of humanitarian organizations, and for the purpose of clarification, they are described as follows. The following broad categorical definitions are widely accepted among the agencies of the international relief community. These are not definitive categories into which each organization will neatly fit, but they have become part of standardized nomenclature in disaster response:

- *Nongovernmental organization (NGO).* The general term for an organization made up of private citizens, with no affiliation with a government of any nation other than the support from government sources in the form of financial or in-kind contributions. These groups are motivated by greatly varying factors, ranging from religious belief to humanitarian values. NGOs are considered national if they work in one country, international if they are based out of one country but work in more than four countries, and multinational if they have partner organizations in several countries. Oxfam and the ICRC are examples of multinational NGOs. NGOs can be further defined according to their functionality. Examples of these would be the religious groups, such as the Catholic Church; interest groups, such as Rotary International; residents' organizations; occupational organizations; educational organizations, and so on.
- *Private voluntary organization (PVO).* An organization that is nonprofit, tax-exempt, and receives at least a part of its funding from private donor sources. PVOs also receive some degree of voluntary contributions in the form of cash, work, or in-kind gifts. This classification is steadily being grouped together under the more general NGO classification. It should be mentioned that while all PVOs are NGOs, the opposite is not true.
- *International organization (IO).* An organization with global presence and influence. Although both the UN and ICRC are IOs, only the ICRC could be considered an NGO. There exists international law providing a legal framework under which these organizations can function.
- *Donor agencies.* Private, national, or regional organizations whose mission is to provide the financial and material resources for humanitarian relief and subsequent rehabilitation. These donated resources may go to other NGOs, other national governments, or to private citizens. Examples of donor agencies are USAID, the European Community Humanitarian Organization (ECHO), and the World Bank.
- *Coordinating organizations.* Associations of NGOs that coordinate the activities of hundreds of preregistered member organizations to ensure response with maximized impact. They can decrease the amount of overlap and help distribute need to the greatest range of victims. Also, they have the ability to analyze immediate needs assessments and recommend which member organizations would be most effective in response. Examples of coordinating

organizations include InterAction and the International Council for Voluntary Agencies (ICVA).

NGOs bring to the field several resources. First, they are well regarded as information-gathering bodies, and thus are vital in establishing accuracy in the development of damage and needs assessments. They tend to provide a single skill or group of specific technical skills, such as the medical abilities of Medicin sans Frontiers (MSF, Doctors without Borders) or Oxfam's ability to address nutritional needs. The sheer number of helping bodies that are provided by the involvement of NGOs allows for a greater capability to reach a larger population in less time. Finally, the amount of financial support provided as a result of the fundraising abilities of NGOs brings about much greater cash resources to address the needs of victims.

These organizations can be characterized by several commonly seen characteristics:

1. *They value their independence and neutrality.* In situations of civil conflict, being perceived as independent is vital to safety and success because they could become targets if associated with an enemy group, or denied access to victims located in territory under the control of a certain warring faction. For this reason, there is often great reluctance on the part of NGOs to share all information to involved governments, to be seen as assisting one group over another, or to report observed war crimes to international tribunals. This independence is advantageous in situations where one national government does not want to be seen as needing the assistance of another national government but is willing to accept the help of autonomous bodies.
2. *They tend to be decentralized in their organizational structure.* For instance, they tend to work without definitive hierarchy and succeed through greater field-level management.
3. *They are committed.* NGOs are often involved not only in the disaster relief, but also in the long-term recovery efforts that follow for months or years. NGO employees are often so dedicated as to repeatedly put themselves in harm's way to deliver aid to victims.
4. *They are highly practice-oriented.* They tend to improvise in the field as necessary and provide on-site training as part of their regular procedures. They rarely use field guides to direct their work, relying rather on the individual experience of employees and volunteers. (CDMHA)

Perhaps the most well-known and most widely established NGO, the Red Cross, will be discussed as an example.

## The International Red Cross

The International Red Cross/Red Crescent Movement consists of the International Federation of Red Cross and Red Crescent Societies (IFRC) and the International Committee of the Red Cross (ICRC). The concept of the Red Cross was initiated by Henry Dunant in 1859, following a particularly brutal battle in Italy that he witnessed. Dunant gathered a local group to provide care for the battle-wounded through medical assistance, food, and ongoing relief. Upon returning to Switzerland, he began the campaign that led to the International Committee for Relief of the Wounded in 1863, and eventually the ICRC. The Committee, and their symbol of a red cross on a white background, has become the standard of neutral wartime medical care of wounded combatants and civilians.

The IFRC was founded in 1919 and has grown to be the world's largest humanitarian organization. After World War I, American Red Cross War Committee president Henry Davison proposed a creation of a League of Red Cross Societies, so that the expertise of the millions of volunteers from the wartime efforts of the ICRC could be used in a broader scope of peacetime activities. Today, the IFRC includes 195 member societies, a Secretariat in Geneva, and more than 60 additional delegations dispersed throughout the world.

The IFRC conducts complex relief and recovery operations in the aftermath of disasters throughout the world. Their four areas of focus include promoting humanitarian values, disaster response, disaster preparedness, and health and community care. Through their work, they seek to "improve the lives of vulnerable people by mobilizing the power of humanity," as stated in their mission. These people include those who are victims of natural and manmade disasters and postconflict scenarios.

Like the UN, the IFRC is well established in most countries throughout the world and is well poised to assist in the event that disaster strikes. Volunteers are continuously trained and utilized at the most local levels, providing a solid knowledge base before a major need presents itself. Cooperation among groups, through the federation, provides an enormous pool of people and funds from which to draw when local resources are exhausted.

When a disaster strikes and the local capacity is exceeded, an appeal by that country's national chapter is made for support to the Federation's Secretariat. As coordinating body, the Secretariat initiates an international appeal for support to the IFRD and many other outside sources and provides personnel and humanitarian aid supplies from its own stocks. These supplies, which can be shipped in if not locally available, pertain to needs in the areas of health, logistics and water specialists, aid personnel, and relief management.

The appeal for international assistance is made an average of 30 times per year, and these assistance projects can continue for years. Long-term rehabilitation and reconstruction projects, coupled with the goal of sustainable development and increased capacity to handle future disasters, have become the norm in regards to major disasters in the poorer countries. The following is how the IFRC responds to international disasters.

Depending on the complexity of the required response, a Field Assessment and Coordination Team (FACT) may be deployed to assist the local chapter in determining the support needs for the event. The teams, which are deployable to any location with only 24 hours' notice, consist of Red Cross/Red Crescent disaster managers from throughout the IFRC, bringing with them skills in relief, logistics, health, nutrition, public health, epidemiology, water and sanitation, finance, administration, and psychological support. The team works in conjunction with local counterparts and host-government representatives to assess the situation and determine what the IFRC response will consist of. An international appeal is drafted, and then launched, by the Secretariat in Geneva. The teams stay in-country to coordinate the initiation of relief activities. Once the effort has stabilized and has become locally manageable, the FACT concedes its control to the local Red Cross headquarters.

In 1994, following a spate of notably severe disasters (i.e., the Armenian earthquake, the Gulf War Kurdish refugee problem, and the African Great Lakes Region crisis), the IFRC began to develop an Emergency Response Unit (ERU) program to increase disaster response efficiency and efficacy. These ERUs are made up of preestablished supplies, equipment, and personnel, who respond as a quick-response unit on a moment's notice and are trained and prepared to handle a much wider range of scenarios than before. This concept, similar to the UNDP Emergency Response Division (ERD), has already proven effective in making IFRC response faster and better, through several deployments, including Hurricane Mitch in Honduras. The teams, upon completion of their response mission, remained in-country to train the locals in water and sanitation issues, thus further ensuring the sustainability of their efforts. ERU teams are most effective in large-scale, sudden-onset, and remote disasters.

Finally, the IFRC is heavily engaged in disaster preparedness and has identified several strategies toward mitigation they hope to achieve by 2010. These activities, which relate to reducing the impact of disasters whenever possible and to working toward better prediction and prevention methods, are becoming a fundamental component of local Red Cross/Red Crescent Society programs. The IFRC has recognized the following four points of action as most vital:

- Reducing the vulnerability of households and communities in disaster-prone areas and improving their ability to cope with the effects of disasters
- Strengthening the capacities of National Societies in disaster preparedness and post-disaster response
- Determining a role and mandate for National Societies in national disaster plans
- Establishing regional networks of National Societies that will strengthen the Federation's collective impact in disaster preparedness and response at the international level

They plan to increase local capacity to handle disasters, thus decreasing the magnitude of international assistance required on disaster onset. This increase in capacity will eventually result in a decreased loss of life and property, as each country becomes more developed and more able to prevent catastrophe. The IFRC aims to accomplish these results through their regular local capacity-building projects, performed in conjunction with research and analysis, which includes the following:

- Hazard prediction
- Risk and vulnerability assessment of individual groups or regions
- Assessment of local strength and capacity in disaster response
- Response network development
- Assessing of National Society disaster mitigation and response capacity
- Assessing national government preparedness and response plans

According to the Geneva Mandate on Disaster Reduction, which was adopted in 1999, the IFRC declared:

> We shall adopt and implement policy measures at the international, regional, sub-regional, national and local levels aimed at reducing the vulnerability of our societies to both natural and technological hazards through proactive rather than reactive approaches. These measures shall have as main objectives the establishment of hazard-resilient communities and the protection of people from the threat of disasters. They shall also contribute to safeguarding our natural and economic resources, and our social well being and livelihoods.

# ASSISTANCE PROVIDED BY THE U.S. GOVERNMENT

## U.S. Agency for International Development

The United States has several means by which it provides assistance to other nations requiring such aid in the aftermath of a disaster, accident (nuclear, biological, or chemical), or conflict. The U.S. agency tasked with providing development aid to other countries, the U.S. Agency for International Development (USAID), has also been tasked with coordinating the U.S. response to international disasters. USAID was created in 1961 through the Foreign Assistance Act, which was drafted to organize U.S. foreign assistance programs and separate military and nonmilitary assistance. One branch of USAID, the Bureau for Humanitarian Response (BHR), manages the various mechanisms with which the United States can respond to humanitarian emergencies of all types. The office under BHR that most specifically addresses the needs of disaster and crisis victims by coordinating all nonfood aid provided by the government is the Office of U.S. Foreign Disaster Assistance (OFDA).

## Office of Foreign Disaster Assistance

The OFDA is divided into four distinct subunits: Disaster Response Division (DRD), Prevention, Mitigation, Preparedness and Planning (PMPP), Operations Support (OS), and Program Support (PS). The DRD handles the U.S. assistance provided to foreign disasters. The PMPP assists foreign nations with assistance to develop their ability to mitigate and prepare for disasters. The OS division handles the technical and logistical support of all OFDA projects, and the PS division works with the OFDA financial and accounting systems.

The administrator of USAID holds the title of President's Special Coordinator for International Disaster Assistance. When a disaster is declared in a foreign nation by the resident U.S. ambassador (or by the Department of State, if one does not exist), the USAID administrator is appealed to for help. This can be done when the magnitude of the disaster has overwhelmed a country's local response mechanisms, the government has requested assistance or will at least accept it, and it is in the interest of the U.S. government to assist. The OFDA is authorized to immediately disburse $25,000 in emergency aid to the U.S. Embassy to be spent at the discretion of the ambassador for immediate relief. The OFDA also can immediately send regional

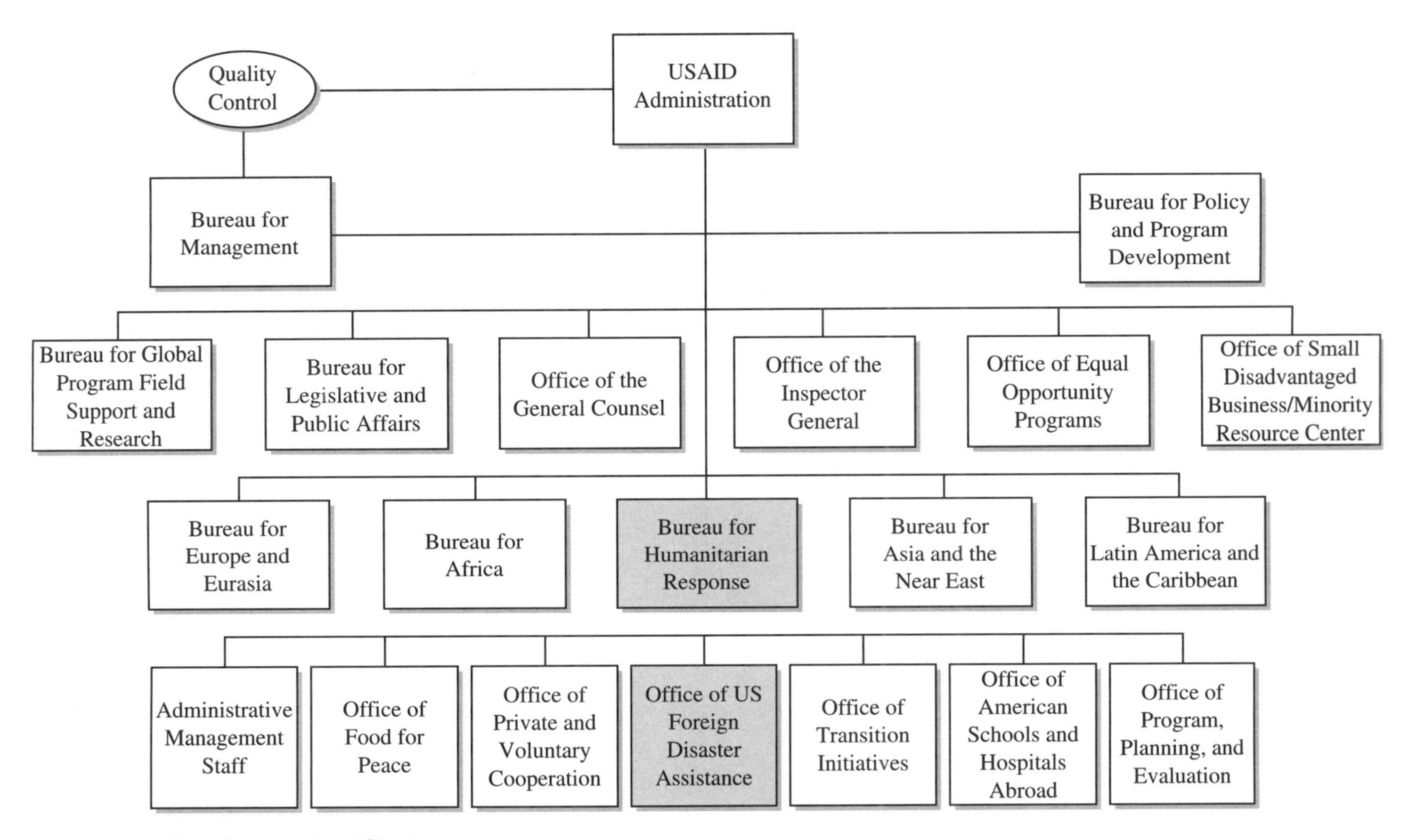


**Figure 8-1** USAID Organizational Chart.

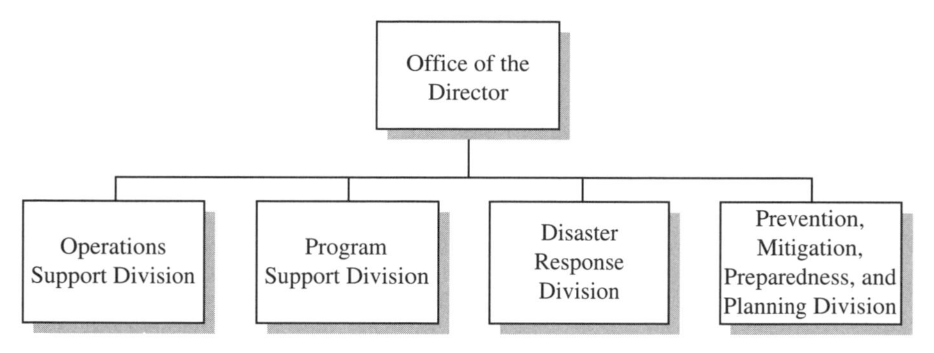


**Figure 8-2** OFDA Organizational Chart.

advisors with temporary shelter and medical aid supplies from one of four OFDA stockpiles in Guam, Italy, Honduras, and the United States.

If the disaster is considerable in size, a Disaster Assistance Response Team (DART) is deployed to the country to assess the damages and recommend the level of assistance that should be made by the U.S. government. DARTs work quickly to develop a strategy to coordinate U.S. relief supplies; provide operational support; coordinate with other donor countries, UN agencies, NGOs, and the host government; and monitor and evaluate projects carried out with U.S. funds. In the largest of disasters, Response Management Teams (RMTs) may be established in both Washington, D.C., and the disaster site, to coordinate and offer administrative assistance and communication for the several DARTs that would be deployed.

The OFDA recently developed a Technical Assistance Group (TAG) to increase its capabilities in planning and programming. TAGs consist of scientists and specialists in agriculture and food security, emergency and public health, water and sanitation, geoscience, climate, urban planning, contingency planning, cartography, and so on. TAGs work with DARTS and RMTs in response, as well as USAID development missions in preparation and mitigation for future disasters.

In addition to the direct aid and logistical and operational support offered, the OFDA provides grants for relief assistance projects. These projects are primarily carried out by PVOs and NGOs, as well as IOs, the UN, and other various organizations (such as a pilots' club that is hired to transport supplies). Not all of this monetary aid goes to response, however. The PMPP works to facilitate projects that aim to reduce the impact of disasters before they happen again. These types of projects seek to empower national governments to make them less likely to need international assistance in subsequent events. All of these organizations are carefully monitored by the OFDA to ensure that they are working efficiently and are spending monetary resources sensibly.

## Other USAID Divisions

Under the USAID BHR, several other offices provide humanitarian aid. The Office of Food for Peace (FFP) handles all of the U.S. government's food assistance projects (U.S. food aid is categorized as Title II or Title III, with the first having no repayment obligations, and the second considered a bilateral loan). The Office of Transition Initiatives (OTI) works in postconflict situations to help sustain peace and establish democracy. The Department of State Bureau for Population, Refugees and Migration (PRM) provides monetary grants to NGOs, PVOs, IOs, and the UN to respond to emergency refugee emergencies. A good portion of this assistance goes directly to the UNHCR. Lastly, the Department of Defense (DOD) responds through their Office of Peacekeeping and Humanitarian Affairs (PK/HA). It is important to note that the developed nations of the world are highly unlikely to receive U.S. assistance on the level that is provided to the developing nations.

## The U.S. Military

The U.S. Military is often involved in relief efforts of natural and technological disasters and CHEs. The involvement of the military, a well-funded and equipped force whose primary function is national defense, brings about an entirely new perspective to the area of operations. It is often argued that nobody is better equipped to handle disasters than the military, with their wide assortment of heavy equipment, enormous reserve of trained personnel, and common culture of discipline and mission-oriented standard operation; however, it is also said that the military is a war agency, not a humanitarian assistance agency, and that these two organizational ideals are too fundamentally and diametrically opposed in practice to allow for effective military involvement.

The assistance of the military is normally requested by USAID/OFDA through the DOD Office of Political/Military Affairs. The chain of command for military operations begins with the President of the United States and the Secretary of Defense, collectively referred to as the National Command Authority (NCA). The NCA, which directs all functions of the U.S. Military, is advised by the Joint Chiefs of Staff (JCS) of the Army, Navy, Air Force, and Marines. The entire military force is divided up into five geographic Areas of Responsibility (AORs) and two functional commands, as follows:

- U.S. Atlantic Command (USACOM): Norfolk, VA, headquarters
- European Command (EUCOM): Stuttgart, Germany, headquarters
- Pacific Command (PACOM): Honolulu, HI, headquarters
- Central Command (CENTCOM): Tampa, FL, headquarters
- Southern Command (SOUTHCOM): Miami, FL, headquarters
- Special Operations Command (SOCOM): In command of special operations, including the Special Forces, Civil Affairs, and Psychological Operations, Tampa, FL, headquarters
- Transportation Command (TRANSCOM): Provides management for all air/sea/land transportation, Scott Air Force Base, IL, headquarters

The U.S. Military is heavily involved in the response to international disasters through organized operations termed Foreign Humanitarian Assistance (FHA) or Humanitarian Assistance Operations (HAO). FHAs are authorized by the DOD Office of Political/Military Affairs (DODPM) at the request of the OFDA (the President, as Commander-in-Chief, gives final authorization for any support operation). Assistance may be provided in the form of physical or technical support, such as logistics, transportation, communications, relief distribution, security, and emergency medicine. In emergencies of natural or manmade origin that do not involve conflict, the role of the military is to provide support, rather than leadership, to the national government and the overall relief community.

The military is known for its self-contained operational abilities, arriving on-scene with everything they need, so to speak. Usually, they provide more than adequate personnel and supplies for the mission they were called to act upon. Once in-country, they work under the strict guidelines of Force Protection (enforced security of all military and civilian personnel, equipment, and facilities associated with their mission) and Rules of Engagement (ROE, a structured, preestablished guideline of "circumstances and limitations under which the military will initiate or continue combat engagement"). The ROE dictate military action in both peacekeeping and disaster operations.

If a particular command unit is tasked with assisting a relief operation, they may deploy a Humanitarian Assistance Survey Team (HAST) to conduct a needs assessment, which relates to the specific functions the military is suited to address. These assessments are occasionally much different than those generated by more humanitarian-based organizations, such as the UN or OFDA, because the military operates in such a fundamentally different fashion. The concerns of the HAST tend to focus on the military support requirements and the logistical factors involving deployment of troops. A Joint Task Force (JTF) will be established soon after to handle the management and coordination of military personnel activities, with a Commander for the JTF designated as the person in charge of the operation on-site; however, if an operation involves only one military service, or is minimal in size, a JTF may not be needed.

One of the main roles of the JTF is to establish a Civil Military Operations Center (CMOC). This center effectively functions to coordinate the military support capabilities in relation to the overall response structure involving all other players involved. The CMOC mobilizes requests for assistance from OFDA, the UN, NGOs, and the host government. All intermilitary planning is conducted through this center, including those operations involving cargo transportation and food logistics. This center is the primary node of information exchange to and from the JTF. CMOCs have taken on expanded responsibility in the past, including the reestablishment of government and civil society and the repair or rehabilitation of critical infrastructure.

## THE INTERNATIONAL FINANCIAL INSTITUTIONS

The international financial institutions (IFIs) provide loans for development and financial cooperation throughout the world. They exist to ensure financial and market stability and to increase political balance. These institutions are made up of Member States, arranged on a global or regional basis, which work together to provide financial services to national governments through direct loans or projects. In the aftermath of disasters, it is common for nations with low capital reserve to request increased or additional emergency loans to fund the expensive task of reconstruction and rehabilitation. Without these IFIs, most developing nations would have no means with which to recover. The largest of these IFIs, The World Bank, and one of its subsidiaries, the International Monetary Fund (IMF), are detailed as follows. Other regional IFIs with similar functions include the Inter-American Development Bank (IDB), which works primarily in Central and South America, and the Asian Development Bank (ADB), based in Manila, Philippines, which works throughout the Asian continent.

## The World Bank

The World Bank was conceived in 1944, during World War II, at Bretton Woods. Its inceptive purpose was to rebuild Europe, and France received the first World Bank loan of $250 million in 1947 for postwar reconstruction. This first lending arrangement reflects the standard of World Bank funding, and financial assistance for reconstruction has been provided regularly since that time in response to countless natural disasters and humanitarian emergencies. Most World Bank loans are provided to the Member States that are less developed.

Today, the World Bank is regarded as one of the largest sources of development assistance, and in the 2001 fiscal year, more than $17.3 billion in loans were provided in more than 100 countries. The World Bank is collectively owned by more than 180 countries and is based in Washington, D.C. It comprises several institutions referred to as the World Bank Group (WBG), which includes the International Bank for Reconstruction and Development (IBRD), the International Development Association (IDA), the International Finance Corporation (IFC), the Multilateral Investment Guarantee Agency (MIGA), and the International Centre for Settlement of Investment Disputes (ICSID). Its overall goal is to reduce poverty, and specifically to "individually help each developing country onto a path of stable, sustainable, and equitable growth, [focusing on] helping the poorest people and the poorest countries" (The World Bank). As disasters and CHEs are taking greater and greater toll on the economic stability of so many countries struggling financially, the Bank, as it is often referred to, is taking on a more central role in mitigation and reconstruction.

Developing nations, which are more likely to have no established mitigation or preparedness and therefore have little or no affordable access to disaster insurance, often sustain damage that is considered a total financial loss. In the periods of rehabilitation that follow the disaster, loans are essential to the success of programs and vital if any level of sustainability or increased disaster resistance is to be achieved. There are several points along this cycle in which the Bank lends assistance.

First, in regular financial assistance, the Bank has worked to ensure that borrowed funds are applied toward projects that give mitigation a central role during the planning phase. They utilize their privilege as financial advisor to guide financial planners who may forego these important measures to stretch the loaned capital as far as possible. They work to increase systems of prediction and risk analysis through their projects to adequately develop according to standards that account for recurrent disasters.

Once a disaster occurs, the Bank may be called on for help. Because it is not a relief agency, the Bank will not take on any role in the initial response; however, immediately upon agreeing to participate, it begins work on restoring damaged and destroyed infrastructure and restarting production capabilities. First, a team may be provided to assist in performing initial impact assessments, including an estimate of pure financial losses resulting from the disaster and an estimated cost of reconstruction including raised mitigation standards. Second, it could restructure the country's existing loan portfolio with the Bank in order to allow for expanded recovery projects. Third, projects that have not yet been approved (but are in the application process) can be redesigned to account for changes caused by the disaster. Finally, an Emergency Recovery Loan (ERL) can be granted, which would specifically address the issues of recovery and reconstruction.

ERLs are granted to restore affected economic and social institutions and to reconstruct physical assets such as essential infrastructure. It is important to note that ERLs are not designed for relief activities. They are most appropriate for disasters that have great adverse impact on the economy, are infrequent in nature (as recurrent disasters are accommodated by regular lending schemes), and create urgent needs. The loan is expected to eventually produce economic benefits to the borrowing government. The ERLs are usually implemented within three years and are flexible to accommodate for the specific needs of each unique scenario. Construction performed with the ERL must use disaster-resistant standards and include appropriate mitigation measures, thus providing an overall preparedness for the country affected. Once an ERL has been granted, the Bank coordinates with the IMF, the UNDP, NGOs, and several other international and local agencies to create a strategy that best utilizes granted funds in relation to the reconstruction effort as a whole.

The Bank works to increase resistance to repeat natural disasters through planning support in their regular lending programs that can be aimed toward mitigation and preparedness. The nature of their mission—to alleviate poverty—is in itself a mitigation measure. As part of its lending process, the Bank conducts vulnerability and risk assessments, which necessitate the subsequent consideration of any findings in planning of future development with Bank loans. The Bank is also a source of information on current hazard-resistant technology and provides the expertise for establishing more effective building codes and their enforcement. As countries develop, they increase their capacity to prepare and respond to disasters and establish the legal and political institutions that guide construction and settlement practices that ensure greater overall resilience. The Bank is arguably the most important player in attaining the means to do so.

## The International Monetary Fund

The International Monetary Fund (IMF) was established in 1946 and has grown to a current membership of 183 countries. Its goals are to promote international monetary cooperation, exchange stability, and orderly exchange arrangements; to foster economic growth and high levels of employment; and to provide temporary financial assistance to countries to help ease balance of payments adjustment. It carries out these functions using loans, monitoring, and technical assistance.

In the event of an international disaster or CHE in a member country, the IMF utilizes its Emergency Assistance Specific Facility to provide rapid financial assistance. In these situations, it is not uncommon for a country to have severely exhausted its monetary reserves. The IMF's goals are to rebuild government capacity and to return stability to the local economy. In the event of a natural disaster, funding is directed toward local recovery efforts and for any economic adjustment that may be needed. If the situation is a postconflict one, its aim is to "reestablish macroeconomic stability and the basis for long-term sustainable growth" (IMF). The IMF will only lend assistance if a stable governing body is in place that has the capacity for planning and policy implementation and can ensure the safety of IMF resources. After stability has been sufficiently restored, increased financial assistance is offered, which will be used to develop the country in its postemergency status.

When a country wishes to request emergency assistance, it must submit a detailed plan for economic reconstruction and ensure that it will not create trade restrictions or intensify exchange. If the country is already working under an IMF loan, then assistance can come in the form of a reorganization within existing arrangements. Separate emergency assistance loans are also offered, which do not involve the regular criteria under which the countries must normally operate. These loans, while normally available only up to 25 percent of a country's preestablished lending quota, have been created in quantities reaching 50 percent of quota; however, this funding is provided only when the member country is "cooperating with the IMF to find a solution to its economic problems." These loans are required to be repaid within five years.

A country often requires technical assistance or policy advice because it is in a situation for which it has no experience or expertise. This is common in postconflict situations where a new government has been established and partnerships are being created for the first time. The IMF offers assistance in building capacity to implement macroeconomic policy. This can include tax and government expenditure capacity, the reorganization of fiscal, monetary, and exchange institutions, and guidance in the use of aid resources.

# CONCLUSION

As global populations converge into more concentrated urban settlements, their collective hazard risks amplify. Loss of life and property caused by the realization of these hazard risks will overwhelm the response and recovery capacities of individual sovereign nations to an ever-increasing degree. Many of these disasters, particularly in the lesser-developed nations, will contribute to existing development obstacles and regional instability unless trends toward increased multilateral cooperation in disaster assistance are recognized more widely for their importance. The capabilities and organizational capacities of the international disaster management agencies listed in this chapter, namely national governments, nonprofit organizations, international organizations, and the international financial institutions, are vital for both the preparation and mitigation of hazard risks, and the response and recovery of actualized disasters.

## CASE STUDY: THE GUJARAT, INDIA, EARTHQUAKE

In Calcutta, India, as citizens were just starting to celebrate their country's 52nd Republic Day, highrise apartment buildings began to shake at a barely perceptible intensity. Little did anybody in that city know, they were not experiencing a weak local tremor but the far-reaching effects of the second most deadly earthquake to hit the country in recorded history taking place more than *1,900 kilometers away* in the state of Gujarat. In fact, the massive temblor, which struck at 8:46 a.m. on January 26, 2001, was also felt in Pakistan and Nepal.[1] This event, the worst earthquake to hit the state of Gujarat in 200 years and the most devastating disaster to hit the country of India in the past 50, struck an unprepared nation.

This case study discusses the origins and disaster history for the affected region and the damage inflicted by the Gujarat earthquake (also referred to as the Bhuj earthquake because of the epicenter's proximity to that city). Also examined are the response that followed by institutions including the national government of India and the state government of Gujarat, the government of the United States, the United Nations, and the multilateral lending institutions. Three nonprofits, the Red Cross, CARE, and Catholic Relief Services, are discussed in relation to their assistance, as a sample of the hundreds of agencies that responded.

### The Earthquake

Origins, Geology, Disaster History

Gujarat's location in the west of India, bordering Pakistan, lies within the Himalayan collision zone where two surface plates (the Indo-Australian and the Eurasian) are slowly crashing together to form the world's youngest and tallest mountain chain at a pace of about two centimeters per year.[2] This movement is but one peril in a land that faces many natural challenges.

Cyclones, floods, drought, and earthquakes characterize Gujarat's history.[3] In the past 25 years, more than 3,000 people and 350 livestock have been killed and more than 1 million houses destroyed by almost yearly cyclones. Floods inundate an average of 300,000 hectares of land, damage an average of 37,000 houses, kill 135 people, and affect 2 million human lives in each average one-year span.[4] The district of Kuchchh, which is the largest in the state, is surrounded by a peculiar swamp called the Rann of Kuchchh, which floods annually and isolates the region from the rest of the Gujarat.[5] Drought is almost a yearly occurrence, with a particularly long three-year drought, which led up to and further complicated events discussed in this case.[6]

In terms of earthquakes, there have been many, with incidents measuring over 6.0 or greater on the Richter scale occurring in 1819 (8.3), 1903 (6.0), 1940 (6.0), and 1956 (7.0). Although the high vulnerability to these disasters has been long established as fact, there was no formalized government management plan to mitigate, prepare, or respond when the Gujarat quake struck. As a result, they were totally unprepared to handle the mass casualty events that ensued.[7] Ironically, this earthquake struck surprisingly close in location to the one that had occurred in 1819 along the same fault line in which many fewer lives were lost. A dramatic increase in development in that region with little or no building code enforcement is blamed for the much higher level of casualty even with a lower intensity of shaking.

### Scope of the Quake

This was the largest earthquake to occur in India since an 8.5 magnitude event hit the state of Assam in 1950.[8] The Indian Meteorological Department (IMD) has recorded a Richter magnitude of 6.9 with location being northeast of Bhuj, while the U.S. Geological Survey (USGS) maintains that the magnitude was 7.9, and the epicenter lay north of Bachau in a location 50 kilometers from the IMD site.[9] The depth of the earthquake, also disputed, was eventually confirmed as approximately 20 kilometers, and resulting aftershocks with an unusual depth of 30 kilometers give the impression that the earthquake may have severed the lithosphere.[10] There was little surface deformation because of the depth, with no clearly discernible cracks on the surface such as those seen with more shallow quakes; however, the liquefaction phenomenon was widespread because of the intensity,[11] and in some cases, rivers that had been dry for more than a century became activated.[12]

Most of the communication infrastructure was immediately destroyed, and a good portion of the transportation infrastructure was damaged. The local government had no immediate means to alert the central government of their imminent needs. This resulted in the lack of an initial assessment, and urban search-and-rescue teams were not sent in time to be fully effective in their missions. The bulk of the initial rescue missions were carried out by neighbors helping neighbors, digging with their bare hands and personal tools.[13] Nobody outside the state could have guessed the magnitude of damage they would find in the coming days, and the character of the first response reflected this knowledge gap; however, when the rescue teams reached the relatively easily accessible city of Ahmedabad and observed the damage, they immediately knew they were going to confront worse conditions in Kuchchh, where the epicenter was located. They moved relief material and volunteers to that region without preassessment.[14]

The earthquake caused damage in 7,904 villages in 21 of the state's 25 districts.[15] The district of Kuchchh sustained the bulk of the damage, with more than 400 villages affected. The towns that suffered most significantly were Bhuj, Bachhau, Anjar, Rapar, and Gandhidham, where virtually 100 percent of the buildings were damaged.[16] This district sustained 90 percent of the deaths and 78 percent of the injuries reported overall, and contained 257,000 of the houses damaged or destroyed.[17] Three hundred kilometers from the epicenter, however, in the city of Ahmedabad, 179 buildings were destroyed.[18]

In many of the areas that were isolated, there was no food or medical relief for up to five days, and people began looting what they could in desperation.[19] In Bachhau, where 30,000 people of 40,000 were cut off from the relief, armed gangs formed and began attacking survivors for money or food.[20] These problems ceased almost immediately upon the arrival of assistance, illustrating the effect a timely response can have on the security of an affected region.

Damage Caused

The damage resulting from this earthquake is a good indicator of the extent to which megahazards will affect nations financially in the 21st century because sustained losses repeatedly exceeded $1 billion. The following list summarizes these damages:

- In pure asset losses, the World Bank and Asian Development Bank estimate that India's losses will exceed $2.1 billion.
- The official government death toll, based on family registration of death and most likely severely underestimated as a result, is 20,005 people; 166,812 were injured, about 20,000 seriously.[21]
- Almost 16 million people, or 1 in 3 in the state of Gujarat, were affected in some way by the January 26 events.[22]
- About 400,000 structures collapsed, and an additional 500,000 to 800,000 were damaged. In the Kuchchh district alone, 300 primary healthcare centers and 1,300 child nutrition centers were lost.

The damage to the state's infrastructure, administration, and communications was extensive and remained a major burden on resources in the reconstruction phase.[23] Several of the sustained damages are listed below:

- The main telecommunications link with Kuchchh snapped and 147 exchanges were damaged in the initial tremor, confounded further by 82,000 damaged phone lines.[24] The remaining open lines were quickly flooded.
- Most power facilities were damaged to some extent, and 925 villages lost power.[25]
- Drinking water and irrigation systems were affected in 1,340 villages, with 1,100 of those villages reporting severe damages.
- Of 240 damaged reservoir dams that supply the water for these irrigation and domestic needs, 20 need to be completely rebuilt.[26]
- More than 100 kilometers of roads were severely damaged, several railroad lines needed repair, and 5 of 10 piers at Kandla Port (the major shipping port in the state) were destroyed.[27]
- Approximately 9,600 primary schools, 2,040 secondary schools, and 140 technical institutions will need to be rebuilt.[28]
- The handicraft industry in Kuchchh suffered the loss of more than 3,000 artisans, and in Dhamadka village, almost 70 percent of the workers in this industry were lost. More than 3,000 small-scale and cottage industries and 20 medium to large-scale enterprises were affected.[29]

Because the impact of this event was not initially communicated to the Government of India, a resulting underestimation of its severity was conveyed to the world community of responders.[30] Much of the initial response was further hampered by the fact that many of the responders (e.g., fire, police, health) were either dead, injured, or attending to family emergencies, which diverted their attention away from the greater relief effort.[31] The scope of rehabilitation required is close to inconceivable, and 18 months later many anxious people were still sleeping out in the open or under plastic sheeting.

### The Participants

The State- and National-Level Governments of India

This event was "the biggest challenge Gujarat has ever faced."[32] By most accounts, the initial response by the state government was nonexistent, primarily because of the complete lack of emergency preparedness and resultant chaos that ensued. Police and fire brigades, the personnel that traditionally respond first in these situations, were occupied with duties related to security and logistics for flag-raising ceremonies and parades.[33] Most government personnel were taking advantage of the long weekend and were not prepared to suddenly return to work.[34]

The Government of Gujarat immediately airlifted a team of five officials headed by the Additional Chief Secretary, which arrived in Bhuj within six hours of the first tremors.[35] This team, although experienced in the management of engineering and medical response, was much too small to handle an event of this magnitude. To increase the rescue staff available, all government officials were officially called off vacation, and an appeal for volunteerism was made to doctors, engineers, retired government officials, and others with applicable skills.[36] Schools and colleges were uniformly closed to ensure that students would be available for the relief and rescue efforts.[37] A state control room was made functional on the first day, and its effectiveness increased once the communication lines were repaired on day two.[38] Ham radio, satellite, and cellphone stations were established for both public and private use.[39] It was not until the third day, however, that the state government diverted heavy equipment used for irrigation, roads, and construction to the search, rescue, and food distribution operations.[40]

The Government of India, on the other hand, took charge almost immediately and responded to an event that would have challenged even the most developed nations. It is important to note that while Prime Minister Vajpayee never formally requested international assistance, he did let it be known that offers of aid were welcome and would gladly be accepted.[41]

Because of the communication infrastructure problems, the initial Government of India response was small and mounted only in Ahmedabad, where reports of damage could be broadcast.[42] The Government of India had no formal disaster management plan that defined the responsibilities of the separate government agencies, so the approach was centralized. Assets had not been inventoried, and their mobilization was not as rapid as it could have been.[43] Other than these initial issues, the government response was one to be commended.

The Krishi Control Room was set up to coordinate the central government response and provide constant communications and updates.[44] The Chief Secretary began holding twice-daily meetings to review the progress and planning of the relief efforts, and a hotline was set up between the Prime Minister and the State Governor to facilitate communication.[45] Local Emergency Operation Control Rooms of varying capability and equipment were set up in tents or structures that had not collapsed, in the localities that suffered the worst damage. These centers acted as information nodes and assisted in the central government coordination to the sites.[46] Two major locations were established as collection, tracking, and distribution centers at Gujarat College and at a town hall, for the tremendous flow of donated goods.[47]

Doctors and nurses were sent to each region with appropriate medical equipment and vehicles.[48] Fifteen thousand Gujarat Electricity Board personnel and 30 truckloads of equipment were dispatched to repair the electrical power in the affected regions.[49] A government survey team was created and examined the status of the buildings that remained standing to determine their safety.[50] Fifteen thousand Indian military service personnel and significant heavy equipment were deployed to provide transportation and distribution support to relief operations, and to repair the airports and bridges that had been damaged.[51,52] The government sent out a request to businesses that operate cranes, gas cutters, and construction equipment to volunteer their services.[53]

When the temperature began to fall at night, temporary shelters were provided as quickly as possible.[54] The water supply, which was already deficient because of the drought, was supplemented by tankers in Kuchchh. Various foods and cooking supplies were distributed, including the allotment of 20 kilograms of wheat and 5 kilograms of rice for each family. For the many families who lost food ration cards (prequake government subsidies), replacements were given.[55] One month's worth of grass was distributed to cattle owners in the region's hardest hit areas.[56] Public service announcements were taped and announced on radio and television instructing people not to enter damaged buildings that could collapse.[57] Customs and excise taxes on all goods imported or manufactured for the relief efforts were waived, and the ban on foreign technology and foreign aid that was in effect was suspended as well.[58] To show government support and sympathy, Prime Minister Vajpayee visited the area and is said to have foregone regular security and stayed longer than originally planned to convey his message.[59]

The Empowered Group of Central Ministers was created to coordinate the emergency response and met for the first time on January 30, 2001. It consisted of representatives from the departments of Home Affairs, Railways, Textiles, Consumer Affairs, Information & Broadcasting, Defense, Finance, Civil Supplies, Health, Rural Development, Housing, Agriculture, Communications, and Power.[60] Their purpose was to do the following:

1. Consider the report of the Crisis Management Group and give such directions as considered appropriate.
2. Decide on all action necessary to provide immediate relief to the victims.
3. Consider measures necessary for relief and rehabilitation of the affected.
4. Consider long-term institutional and organizational measures that are necessary for management and mitigation of such natural disasters.[61]

For the restoration efforts, the Gujarat Earthquake Rehabilitation Fund was created to raise money. Grants were distributed according to the extent of financial and physical damage. The Government of Gujarat State Disaster Management Authority was created to better enable these reconstruction efforts, heralding $1 billion in aid to assist more than 300,000 families according to level of village damage, distance from the epicenter, and the original house value.[62] A national Department of Earthquake Relief was also created, as part of the department of General Administration.[63] Finally, a plea was made to ban all public celebrations until February 28, and to ask that "those celebrating marriage and other social programs [are] modest and austere."[64]

The U.S. Government

The U.S. Government, one of the largest donors in the relief effort, provided aid through the Department of Defense (DOD) and the U.S. Agency for International Development (USAID) Office of Foreign Disaster Assistance (OFDA). Between the two agencies, the United States contributed $13.1 million to the response effort. The DOD provided airlifts for all of the donated goods, a 2.5-ton truck, two forklifts, two 400-gallon tankers, 10,000 blankets, 1,500 sleeping bags, and 92 50-person tents. A six-person military assessment team consisting of experts in communications, logistics, and technical support, was provided to advise the government responders.[65] The OFDA provided assistance through donated commodities and through grants via organizations such as CARE, Catholic Relief Services, and the World Food Program. Three airlifts by OFDA (valued at $2,426,463), which carried technical equipment, shelters, blankets, sleeping bags, water and sanitation equipment, and other goods, supplied relief to more than 450,000 people.[66] Grant programs that OFDA dollars facilitated included water sanitation, disease surveillance, emergency shelters, relief distribution, medical support, trauma counseling, and food assistance. In addition to these projects, a USAID Disaster Assistance Response Team (DART) of 11 people was dispatched to conduct emergency needs assessments and coordinate the distribution of all relief supplies donated by the U.S. Government.[67] Finally, $100,000 was given to the Prime Minister's Gujarat Rehabilitation Relief Fund.

The United States has remained active in the recovery and rehabilitation of Gujarat. USAID developed the Gujarat Earthquake Recovery Initiative, which is aimed at families in the poorest communities. An allocation of $10 million has been granted, and the funds will come from existing USAID budget resources to be used by various NGOs and multilateral organizations like the UN. There are four established areas for which the funds will be used:

1. Cash, for work and other NGO programs to help repair roads, wells, water systems, homes, workplaces, and other infrastructure needed to restart economic activities
2. Cash, for work and other NGO programs to clear away debris and rubble and repair public facilities such as health clinics and child nutrition centers
3. Survey support, to assess damaged (but still standing) buildings, to determine whether they can be repaired and retrofitted or if they need to be demolished and rebuilt
4. Support, to municipalities and local NGOs to develop community renewal plans that will help reconstruct devastated communities[68]

The United States was just one of many countries that responded, providing a total of about $90 million to support the relief and reconstruction of Gujarat.[69] Other nations that assisted include Australia, Austria, Belgium, Canada, Denmark, Finland,

France, Germany, Greece, Hungary, Ireland, Israel, Italy, Japan, Luxembourg, Monaco, Nepal, the Netherlands, New Zealand, Norway, Oman, Poland, Russia, Singapore, Spain, Sweden, Switzerland, Taiwan, Turkey, and the UK.[70]

The Nongovernmental Organizations

More than 200 NGOs engaged in the response and relief effort in India, creating a daunting task of cooperation and coordination.[71] Initially, there was no built-in government mechanism to organize the relief. Under these chaotic circumstances, the organizations worked out a system of coordination on their own, which attempted to create an optimal working arrangement for the disaster and increase the effectiveness of response to the greatest number of those in need. It is reported that this was the first time coordination efforts such as these had taken place in India, and they were primarily successful.[72] Three of these organizations' responses, those of CARE, Catholic Relief Services, and the Red Cross, are described as follows.

**Care.** CARE mobilized the morning after the quake to perform an initial assessment of the Kuchchh district. They provided an immediate supply of medical equipment, food, blankets, tarps, tents (10,000 family-size), and water-purification tablets. CARE emergency medical teams provided treatment and trauma counseling to survivors in the hard-to-reach areas of Anjar, Bachau, Rapar, and Bhuj.[73] With the help of a USAID grant, they were able to provide food and survival kits to assist 50,000 people, to encourage them to remain in their home areas rather than become displaced.[74]

CARE's work in India lasted through the end of February, helping more than 175,000 people in the remote villages where they felt need was the greatest.[75] In this time, they helped build at least 118 community service facilities (e.g., schools, health centers, government offices) and 105 water systems (locally managed for sustainability), increased access to employment and training for 6,000 people, and rebuilt damaged irrigation systems and watershed management schemes. Overall, their goal was to increase the general capacity of the earthquake victims through many self-help initiatives.

**Catholic Relief Services.** Catholic Relief Services (CRS), like CARE, was the recipient of a large portion of the USAID grants. In addition, they committed $650,000 in private funds, as well as their Africa-based emergency technical unit and staff from various locations including Bosnia.[76] Initial cash resources were designated for the installation of temporary shelter and to meet the personal hygiene needs of more than 65,000 people in 73 villages. Mental Health units were established to provide trauma counseling for the injured, their families, and the most vulnerable groups (e.g., women, children, lower-caste members, elderly, and minorities).[77] One year later, CRS was still working on follow-up projects to increase the likelihood of program success and are creating village resource maps to maximize the overall target population size.[78]

**American Red Cross.** The American Red Cross is one of the most experienced organizations in responding to international disasters of every type. They were one of the first organizations on the ground in Gujarat, working with a team of 11

American experts trained in logistics, communication, mental health, and family tracing. This team supported the overall International Red Cross team of more than 120 people. The Red Cross distributed almost $2 million in supplies to nearly 100,000 victims. Included in this aid were 13,000 five-gallon buckets, 550 rolls of plastic sheeting, 15,000 kitchen sets, 25,000 tarps, 15,000 blankets, and 5,000 tents. They purchased and distributed emergency health kits, from the World Health Organization, which included medicine, intravenous fluids, surgical tools, and other medical supplies.[79]

The Red Cross plans to assist the state of Gujarat in the reconstruction as well. Their current projects aim to do the following:

1. Help rebuild community infrastructure to provide safe, clean water, including the repair and installation of water collection, storage, and sanitation.
2. Develop a trained network of Indian mental health professionals who will provide mental health counseling for this and other disasters.
3. Provide community health education programs to improve access to basic healthcare and prevent the spread of communicable diseases.[80]

These efforts complement the $15 million in aid provided by the International Federation of Red Cross/Red Crescent Societies (IFRC), of which a portion was used to construct a 310-bed, high-tech emergency hospital in Bhuj.[81,82]

### The United Nations

The UN agencies responded immediately, having access to all government information through their established in-country presence. The UN Development Programme (UNDP) was coordinator, assisting in responses of the World Food Programme (WFP), the UN Children's Fund (UNICEF), the Office of the Coordination of Humanitarian Affairs (OCHA), the International Labor Organization (ILO), and the World Health Organization (WHO). The accomplishments of each of these agencies is described as follows:

**UN Development Programme.** The UNDP deployed its Disaster Response Team, whose responsibility was to coordinate the entire emergency response until the UNDP could formally assume that role.[83] In addition, supported by $2.75 million from the governments of the United States, Britain, and Italy, the UNDP coordinated the UN body needs assessments, activity identification, project proposal design and implementation, monitoring, and quality control.[84] The UNDP and the UN volunteers they oversee worked to address the issue of the houses destroyed in the quake. Using "roaming teams," they worked with local communities to develop and fund projects for the distribution of building materials and the construction of temporary shelters. These teams also monitored the progress of the projects. The UNDP provided $100,000 for immediate relief through a project in partnership with two of the leading women's organizations in Gujarat: the Self-Employed Women's Association (SEWA) and the NGO Kutch Mahila Vikas Sangthan, who put together survival kits for families in addition to helping with general housing issues.[85] The UNDP sent 35 UN volunteers into several regions where no other NGOs had initiated work or provided assistance, and plans to eventually have 5,000 volunteers working on the overall recovery effort.[86]

The UNDP continues to be the UN coordinating body for reconstruction, and this is to be a long and arduous task. They have continued to work with the central Government of India and the state Government of Gujarat in implementing plans to provide permanent housing to the homeless, using construction design that is resistant to the many risks encountered in that region.[87] All of these projects are merely in addition to those the UNDP already is conducting throughout India.

**The World Food Programme.** The WFP launched a $4.14 million project that provided relief food rations to 300,000 people for four months. Most of these people received packages of wheat flour and lentils, to help them survive the months following the earthquake. They specifically targeted a group of 178,000 children below the age of five and pregnant and nursing mothers, and provided them with highly nutritious biscuits and a fortified blended food called Indiamix. A special Joint Logistics Center was initiated in Bhuj on February 11, with a $2.3 million budget, to coordinate the overall relief efforts for the victims and airlift the relief material from a UN Humanitarian Response Depot (UNHRD) in Brindisi, Italy, to Bhuj.[88]

**UN Children's Fund.** Just two days after the earthquake struck, UNICEF sent a team of 15 members based in Gujarat to distribute 15,000 blankets, 1 million chloroquine tablets to purify drinking water, and medical supplies that could help 30,000 people for three months. In the next 72 hours, they provided an additional $600,000 in medical equipment. Over the course of the next few weeks, during the response phase, UNICEF supplied 83 mobile water tankers, countless medical supplies of every type, 75,000 blankets, measles vaccines to more than 400,000 children, water supply systems, 700 large tents (to act as temporary classrooms and healthcare centers), school supplies, vitamin A for 1 million children, 1 million oral rehydration packets, refrigerators, generators, and 106,000 family survival kits.[89] UNICEF is continuing to work with the Government of Gujarat to rebuild many of the schools that were damaged or destroyed, and is helping the communities in the state prepare emergency preparedness plans. UNICEF contributed more than $21 million to relief and reconstruction.

**Office of the Coordination of Humanitarian Affairs.** OCHA sent a five-member UN disaster assessment and coordination team on January 27 to assist the UNDP in the response phase of the disaster.[90] It provided an emergency grant of $150,000 from its own resources and from prepositioned funds from the Governments of Denmark and Norway to purchase tents and blankets. Together with the WFP, OCHA organized the three relief flights from the UNHRD in Brindisi, Italy. Periodically during the response phase, OCHA issued Situation Reports in order to keep the international community informed and to raise support for the affected population.[91]

**International Labor Organization.** The ILO's activities were aimed at creating short-term work opportunities in cleanup, rebuilding the infrastructure and housing, and "protecting vulnerable groups such as young women and children."[92] They established programs that addressed aspects of disaster recovery relating to their main concern of labor issues. These projects sought to gather statistics relating to the effect on the job market from losses in employees and employment, migration flows, and the skills of the victims. Using what they refer to as "labor-intensive methods," they provided immediate employment opportunities to stimulate local markets and provide people with self-reliance. They concentrated on the most vulnerable groups, such as women and children, and worked with other agencies (such as UNICEF) to curb the disaster effects that lead to child labor, child trafficking, and sexual exploitation.[93]

**World Health Organization.** WHO sent a team of nine public health experts to Gujarat to perform a rapid health assessment of the region. A disease surveillance desk was established in the main emergency operations center in Bhuj to monitor the possible outbreak of disease (which often appears in mass-casualty events).[94] Experts from WHO provided technical advice to the state government and health officials on public health issues. They also provided emergency health materials, including trauma kits, emergency health kits, and other essential medical supplies, all within the first days of the disaster. What was most needed, however, was the rehabilitation of the damaged and destroyed healthcare facilities, and they were working with the experience they had acquired in the same region after the 1999 cyclone that caused similar destruction.[95]

### The International Development Banks

It is important to mention the international development banks that worked with the Government of India to finance reconstruction loans that are essential to the recovery of the state. Although these institutions played a vital role in establishing the preliminary and final assessments of the damages and reconstruction needs resulting from the quake, they do not perform any duties related specifically to the response. Their involvement in the reconstruction is essential because they are providing the capital, without which nothing could be rebuilt, and are working with the Government of India in developing a reconstruction plan that will be able to better sustain the types of natural disasters that afflict the area on a regular basis.[96]

## References

1. Government of Gujarat, India, *Earthquake in Gujarat*, (n.d.) http://gujaratindia.com/preliminary.html, retrieved October 30, 2001, p. 1.
2. World Bank and Asian Development Bank, *Gujarat Earthquake Recovery Program: Assessment Report*, March 14, 2001, p. 10.
3. *Ibid*, p. 6.
4. *Ibid*, p. 9.
5. Krishna Vatsa, *The Bhuj Earthquake, District of Kutch, State of Gujarat (India) January 26, 2001*, March 16, 2001, DRM-World Institute for Disaster Risk Management, p. 1.
6. U.S. House of Representatives, *The Earthquake in India: The American Response*, March 1, 2001, Washington D.C., U.S. Government Printing Office, p. 10.

7. EERI, *EERI Preliminary Government Response Report, Earthquake in Gujarat, India,* January 26, 2001, (n.d.) www.eeri.org/earthquakes/reconn/bhuj_india/pgovtrest.html, retrieved October 29, 2001, p. 1.
8. USAID, *India—Earthquake, Fact Sheet #14 (FY 2001),* February 8, 2001, www.usembassy.state.gov/posts/in1/wwwhguj.html, p. 1.
9. Ravi Sinha and Rajib Shaw, *The Bhuj Earthquake of January 26, 2001: Consequences and Future Challenges,* April 26, 2001, p. 1.
10. CIRES, *26 January 2001 Bhuj Earthquake, Gujarat, India*, August 2001, University of Colorado, http://cires.Colorado.edu/~bilham/gujarat2001.html, p. 3.
11. *See* note 2, p. 2.
12. *See* note 10.
13. *See* note 9, p. 4.
14. *Ibid*, p. 11.
15. *Ibid*, p. 4.
16. *See* note 5, p. 6.
17. *See* note 6.
18. *See* note 1.
19. *Ibid*, p. 3.
20. "Food Riots Hamper Quake Relief Work in Gujarat," *Times of India*, January 30, 2001.
21. *See* note 6.
22. Government of India, *The Government of India's Official Website on the Gujarat Earthquake,* (n.d.) http://gujarat-earthquake.gov.in/mainpage.htm, p. 3.
23. *Ibid*, p. 2.
24. *Ibid*, pp. 1–9.
25. *Ibid*, p. 14.
26. *See* note 2, p. 1.
27. *See* note 22, pp. 11–13.
28. *See* note 2, p. 1.
29. *Ibid.*
30. *See* note 1, p. 2.
31. *See* note 22, p. 1.
32. *See* note 1, p. 11.
33. *See* note 7.
34. *See* note 1.
35. *Ibid.*
36. *Ibid*, p. 1.
37. *Ibid*, p. 3.
38. *Ibid*, p. 6.
39. *Ibid*, p. 3.
40. *See* note 6, p. 11.
41. *See* note 9, p. 9.
42. *Ibid.*
43. *See* note 22, p. 5.
44. *See* note 1, pp. 5, 7.
45. *See* note 9, p. 10.
46. *See* note 1, p. 7.
47. *Ibid*, p. 6.
48. *See* note 5, p. 7.
49. *See* note 1, p. 7.
50. *See* note 8, p. 8.
51. *See* note 1, p. 6.
52. *Ibid*, p. 6.
53. *Ibid*, p. 4.
54. *Ibid*, p. 9.

55. *Ibid*, p. 11.
56. *Ibid*, p. 6.
57. *Ibid*, pp. 33, 36.
58. *Ibid*, p. 4.
59. *See* note 22, p. 5.
60. Government of India, *The Central Government's Relief Efforts*, 2001, http://www.gujarat-earthquake.gov.in/final/ministry-wise.htm, last accessed May, 2003.
61. *See* note 5, p. 9.
62. *See* note 1, p. 11.
63. *Ibid.*
64. *See* note 8, p. 3.
65. *Ibid.*
66. *Ibid.*
67. USAID, *Summary of U.S. Government Assistance for the Victims of the Gujarat Earthquake,* February 12, 2001, p. 1.
68. *See* note 6, pp. 14–15.
69. *Ibid*, p. 39.
70. *See* note 8, p. 3.
71. "Gujarat Tragedy: Could Be a Blessing in Disguise," *The Statesman* (India), April 23, 2001, http://web.lexis-nexis.com.
72. *See* note 9, p. 4.
73. *See* note 6, p. 23.
74. USAID, *USAID Awards $2.2 Million in Contracts for Indian Earthquake,* February 5, 2001, USAID Press Office, p. 1.
75. *See* note 6, p. 23.
76. *Ibid*, p. 16.
77. *Ibid*, pp. 13–19.
78. Interaction, *Earthquake in India*, June 2001 www.interaction.org/india00/index.html, p. 6.
79. American Red Cross, *India Quake Relief 2001,* (n.d.) www.redcross.org/news/in/0101bhuj/india.html, p. 2.
80. *Ibid*, p. 3.
81. *See* note 9, p. 12.
82. *See* note 2, p. 3.
83. *Ibid*, p. 1.
84. Madhu Nainan, "Quake-Proof Houses Set the Model for Gujarat Reconstruction," *Agence France-Presse*, March 13, 2001, p. 5.
85. UNDP Press Release, *UN Steps Up Relief Operations in Gujarat,* United Nations, February 3, 2001, p. 2.
86. *Ibid*, p. 6.
87. *See* note 2, p. 45.
88. UN India, *The UN System Response to the Gujarat Earthquake,* August 2, 2001, www.un.org.in/dmt/guj/unresguj80201c.htm, retrieved October 9, 2001, p. 6.
89. UNICEF, *India: Earthquake, Complete Emergency Report,* (n.d.) www.unicefusa.org/alert/emergency/indiaeq/archive.html, p. 6.
90. *See* note 2, p. 3.
91. *See* note 88, p. 7.
92. UNDP Press Release, "UN Steps Up Relief Operations in Gujarat," United Nations, February 3, 2001, p. 2.
93. *See* note 88, p. 7.
94. *See* note 2, p. 3.
95. *See* note 92, p. 2.
96. *See* note 2, p. 4.

# 9. Emergency Management and the New Terrorist Threat

## INTRODUCTION

The terrorist attacks of September 11 prompted dramatic changes in emergency management in the United States. These attacks and the subsequent Anthrax scare in Washington, D.C., in October 2001 have been the impetus for a reexamination of the nation's emergency management system, its priorities, funding, and practices. These changes are ongoing and will continue for the foreseeable future.

Before September 11, the Nunn-Lugar legislation provided the primary authority and focus for domestic federal preparedness activities for terrorism. Several agencies—the Federal Emergency Management Agency (FEMA), Department of Justice (DOJ), Department of Health and Human Services (DHHS), Department of Defense (DOD), and the National Guard—were all involved and jockeying for leadership of the terrorism issue. There were some attempts at coordination but, in general, agencies pursued their own agendas. The biggest difference among the agencies was the level of funding available, with the DOD and DOJ controlling the most funds. State and local governments were confused, felt unprepared, and complained of the need to recognize their vulnerability and needs should an event happen. The TOPOFF exercise held in 1999 reinforced these concerns and vividly demonstrated the problems that could arise in a real event. The events of September 11, unfortunately, validated their concerns and visibly demonstrated the need for changes in the federal approach to terrorism.

The changes fall into five general categories, including (1) first responder practices and protocols, (2) preparing for terrorist acts, (3) funding the war on terrorism, (4) creation of the Department of Homeland Security, and (5) the shift in focus of the nation's emergency management system to the war on terrorism. This chapter explores these categories, identifies issues, and discusses the implications of this new direction for emergency management. Where appropriate, a historic perspective to these changes is provided.

## CHANGES IN EMERGENCY MANAGEMENT AND THE WAR ON TERRORISM

As of the writing of this text in Fall 2002, the first wave of after-action reports on the response to the events of September 11, 2001 at the World Trade Center in New York

City, at the Pentagon in Virginia, and in Washington, D.C., are beginning to be completed and made available to the emergency management community and the public. The principal focus of these after-action reports is on the actions taken by first responders—fire, police, and emergency medical technicians—at the scene of the World Trade Center and the Pentagon. Not surprisingly, these reports identify some basic changes in the practices and protocols of first responders to future terrorist incidents designed to reduce the terrible toll taken on first responders at the World Trade Center. Most of these changes will be implemented at the local level.

Five groups must be fully engaged in the nation's war on terrorism: the diplomats, the intelligence community, the military, law enforcement, and emergency management. The principal goal of the diplomats, intelligence, military, and law enforcement is to reduce, if not eliminate, the possibility of future terrorist attacks on American citizens at home and abroad.

The goal of emergency management should be to be prepared and to reduce the future impacts in terms of loss of life, injuries, property damage, and economic disruption caused by the next terrorist attack. As President Bush and many of his advisors have repeatedly informed the nation, it is not a question of if but rather when the next terrorist attack occurs. Therefore, it is incumbent on emergency managers to apply the same diligence to preparing for the next bombing or biochemical event as they do for the next hurricane or flood or tornado. The focus of emergency management in the war on terrorism must be on reducing the danger to first responders, the public, the business community, the economy, and the American way of life from future terrorist attacks. This change must occur at all levels of the emergency management system: federal, state, and local.

The war on terrorism has resulted in unprecedented funding resources being made available to the emergency management community. For the first time, vast sums of money from the federal government are available for first-responder equipment and training, for planning and exercises, and for the development of new technologies. Funding for FEMA has increased, as has the amount of funds FEMA delivers to state and local emergency management organizations.

Historically, FEMA has distributed about $175 million annually to its state and local emergency management partners. The federal fiscal year 2003 budget contains $3.5 billion for FEMA to distribute to states and local emergency management organizations. This is in addition to funding FEMA received in a supplemental funding bill passed by Congress after the September 11 attacks. New federal funding sources are also opening up for emergency managers from the DOD, the DOJ, and the DHHS to fund contingency plans, technology assessment, and development and bioterror equipment and training. This change in funding for emergency management will be felt most significantly at the state and local levels.

The creation of the Department of Homeland Security (DHS) represents a landmark change for the federal community, especially for emergency management. The consolidation of all federal agencies involved in fighting the war on terrorism follows the same logic that first established FEMA in 1979. At that time, President Carter—at the request and suggestion of the nation's governors—consolidated all of the federal agencies and programs involved in federal disaster relief, preparedness, and mitigation into one single federal agency, FEMA.

The director of the new agency reported directly to the President, as will the DHS Secretary; however, when FEMA is absorbed into the DHS, the FEMA Director will no longer report directly to the President but rather to the DHS Secretary. This change could have a significant impact on FEMA and its state and local partners in managing natural and other technological disasters in the future.

At the request of President George W. Bush, FEMA established the Office of National Preparedness in 2001 to focus attention on the then-undeclared terrorist threat and other national security issues. This was the first step in refocusing FEMA's mission and attention from the all-hazards approach to emergency management embraced by the Clinton administration. The shift in focus was accelerated by the events of September 11 and has been embraced by state and local emergency management operations across the country. A similar shift of focus in FEMA occurred in 1981 at the beginning of the Reagan administration. Then the shift of focus was from disaster management to planning for a nuclear war. For the remaining years of the Reagan administration and the four years of President George H.W. Bush's administration, FEMA resources and personnel focused their attention on ensuring the continuity of government operations in the event of a nuclear attack. Little attention was paid to natural hazard management, and FEMA was left unprepared to deal with a series of catastrophic natural disasters starting with Hurricane Hugo in 1989 and culminating with Hurricane Andrew in 1992. If history repeats itself, the current change in focus away from the all-hazards approach of the 1990s could result in a weakening of FEMA's natural disaster management capabilities in the future.

The remaining sections of this chapter discuss how these changes are affecting emergency management organizations at the federal, state, and local levels. A summary of the response and recovery efforts to the World Trade Center and Pentagon attacks provides a perspective on why such dramatic actions are being proposed.

## SUMMARY OF SEPTEMBER 11 EVENTS

Measuring the far-reaching effects of the events of September 11 on emergency management can be done in a wide variety of ways. In the following sections of this chapter, some of the organizational, funding, technology, and operational changes that these events initiated are discussed. How the focus of emergency management in this country has shifted because of these events is also examined, as well as the size and breadth of these events through an examination of some of the financial costs, principally spending by FEMA and other federal government agencies, in responding to and assisting in the recovery from these events.

When considering the effects of the events of September 11, the first effect that must be considered is the horrific loss of life in New York City, Virginia, and Pennsylvania. As of August 2002, no single tally of the dead in New York City has been agreed on by either official government sources or the media. The New York Police Department lists 2,823 dead and missing, while the city's medical examiner lists 2,819. In the media, the Associated Press lists 2,786 dead, *Newsday* lists 2,814 dead, and *USA Today* lists 2,801 dead.

The attacks on the World Trade Center and the Pentagon together could arguably be considered the first national disaster event, outside of wartime, in the history of the United States. It is the first disaster event in this country that affected all citizens of this country and left all citizens and communities with an uneasy sense of vulnerability; however, because the economic consequences of these attacks were felt in all parts of the United States and, in fact, around the world is what makes this disaster event truly national in scope. Measuring the economic effects of such an event is daunting, and some measures will take years to complete, but a quick review of some of the economic effects measured to date clearly illustrates the breadth and depth of this disaster's effect on the economic well-being of the people of the United States.

## PARTIAL LIST OF ECONOMIC LOSSES CAUSED BY SEPTEMBER 11 EVENTS (AS OF DECEMBER 2001)

### The Airline Industry

The terrorist attacks on the United States have had a major impact on an industry that was already expected to lose between $2.5 and $3.5 billion in the year before September 11. For example, US Airways announced in August that it expected a $160 million loss for the third quarter. Because of this, the airline stated that a restructuring would be necessary if it were to remain competitive. American Airlines also announced that it would have to ground some of its older aircraft because of actual and anticipated losses. In mid-August, Midway Airlines filed for Chapter 11 bankruptcy and laid off about 700 workers.

Following the terrorist attacks, the airline industry was affected by flight curtailments, high overhead costs, and low passenger utilization. Massive layoffs followed, to include Continental (12,000), Delta (13,000), American (11,000), ATA (1,500), and United (11,000), with total expected industry layoffs of 100,000 and projected financial losses of $4.5 billion. Congress then passed an industry assistance package of $15 billion ($5 billion for immediate aid and $10 billion in loans).

Recently, U.S. air travel during the Thanksgiving holiday was estimated to be 1.8 million, as compared to 2.2 million last year. Overall air traffic for October and November of this year is down 25 percent from the same period as last year. U.S. travel abroad is off 30 percent from last year. Since September 11, two major overseas carriers have suffered significant losses. Sabena went into bankruptcy and Swissair received a government bailout to keep afloat but is only working at about half of its usual capacity.

*Sources:* (1) "Airlines—The Unpalatable Truth," Economist.com, November 25, 2001, and (2) "States Most Affected by Airline Troubles," DRI-WEFA Economic Briefings, November 9, 2001.

## The Insurance Industry

The costs to U.S. insurance companies and the reinsurance industry resulting from the terrorist attacks of September 11 are extraordinary. A summary of estimated costs to date are Berkshire Hathaway ($2.3 billion), Lloyd's ($1.8 billion), Munich "Re" ($1.8 billion), Swiss "Re" ($1.4 billion), and Allianz ($.9 billion). Warren Buffet of Berkshire Hathaway indicated that losses of up to $1 trillion could result from acts of terrorism. It is evident that the private insurance market may not be able to absorb such costs.

*Sources:* (1) "Cost to Insurers of U.S. Terrorist Attacks," Economist.com, November 25, 2001, (2) Interview with Lawrence Lindsey, Fox TV News, November 25, 2001, and (3) *New York Times*, November 25, 2001.

## Costs Associated With Federal/State Disaster Assistance

As of November 19, FEMA, the SBA, and the State of New York disaster assistance program for the World Trade Center attacks provided $657 million in relief funds. In terms of individual assistance, more than $164 million has been approved for grants and loans. FEMA approved $17.6 million in disaster housing assistance. The SBA approved $118.7 million in low-interest loans. Also, crisis-counseling grants in excess of $22.7 million were approved. In addition, FEMA funded more than $154 million through other agencies such as the Army Corps of Engineers, the DHHS, and Urban Search and Rescue. Costs to New York City relating to debris removal at Ground Zero were $137.1 million as of November 24. Since September 11, it was estimated that Homeland Security has cost $10 billion.

*Sources:* (1) NBC TV News, November 18, 2001, (2) Fox TV News, November 24, 2001, (3) "Federal/State Disaster Assistance Tops $657 Million," FEMA, *Disasters: Information on Federally Declared Disasters*, November 19, 2001.

## U.S. Unemployment

Since September 11, many service industry workers who rely on the airlines for employment lost their jobs. Food and beverage outlets cut 42,000 workers, and hotels laid off 46,000. An anticipated reduction of 100,000 airline workers will further compound this significant period of unemployment in the United States.

*Source:* "Greatest U.S. Employment Loss in 20 Years," DRI-WEFA Economic Briefings, November 2, 2001.

Another measure of the size of these events is the costs to the federal government of providing disaster relief. As of August 22, 2002, FEMA had approved more than $4.49 billion to the State of New York for emergency and recovery work. This

**Table 9-1** Top Ten Natural Disasters (Ranked By FEMA Relief Costs)

| Event | Year | FEMA Funding* |
|---|---|---|
| Northridge Earthquake (CA) | 1994 | $6.999 billion |
| Hurricane Georges (AL, FL, LA, MS, PR, VI) | 1998 | $2.333 billion |
| Hurricane Andrew (FL, LA) | 1992 | $1.849 billion |
| Hurricane Hugo (NC, SC, PR, VI) | 1989 | $1.308 billion |
| Midwest Floods (IL, IA, KS, MN, MO, NE, ND, SD, WI) | 1993 | $1.141 billion |
| Hurricane Floyd (CT, DE, FL, ME, MD, NH, NJ, NY, NC, PA, SC, VT, VA) | 1999 | $1.085 billion |
| Tropical Storm Allison (FL, LA, MS, PA, TX) | 2001 | $879.5 million |
| Loma Prieta Earthquake (CA) | 1989 | $865.5 million |
| Red River Valley Floods (MN, ND, SD) | 1997 | $734.0 million |
| Hurricane Fran (MD, NC, PA, VA, WV) | 1996 | $621.2 million |

*Amount obligated from the President's Disaster Relief Fund for FEMA's assistance programs, hazard mitigation grants, federal mission assignments, contractual services, and administrative costs as of February 28, 2002. Figures do not include funding provided by other participating federal agencies, such as the disaster loan programs of the Small Business Administration and the Agriculture Department's Farm Service Agency. *Note:* Funding amounts are stated in current dollars.
*Source:* FEMA, www.fema.gov

total represents only FEMA's expenditures on this disaster and does not include expenditures by other federal agencies, insurance companies, and the private sector. According to FEMA records, this total would place the World Trade Center disaster number 2 on FEMA's list of Top Ten Natural Disasters (see Table 9-1; FEMA does not have a comparable list for technological disasters that could be used to compare the events of September 11, so the natural disaster list was used).

Summaries of selected FEMA costs associated with the World Trade Center disaster are presented in Table 9-2.

As of August 21, 2002, more than 90,000 New Yorkers had visited the FEMA-sponsored Applicant Assistance Center in Lower Manhattan. Also, more than 58,000 people seeking FEMA disaster assistance had contacted FEMA on its toll-free registration hotline.

## FIRST RESPONDER EVALUATION

In July and August 2002, two September 11–related after-action reports were released, "Improving NYPD Emergency Preparedness and Response" prepared by McKinsey & Company for the New York City Police Department (NYPD) and "Arlington County After-Action Report on the Response to the September 11 Terrorist Attack on the Pentagon" prepared for Arlington County, Virginia, by Titan Systems Corporation. Both reports are based on hundreds of interviews with event participants and reviews of organizational plans. These reports provide lessons learned and present hundreds of recommendations.

**Table 9-2** Selected FEMA Disaster Expenditures for the World Trade Center Disaster

| Expenditure Type | Total Cost |
|---|---|
| Temporary Housing Assistance | $28.2 million in grants to 5,500 households as of 2/4/02 |
| Mortgage Rental Assistance | $5.9 million as of 2/4/02 |
| U.S. Small Business Administration loans | $257.9 million to 3,200 businesses, homeowners, and renters as of 2/4/02 |
| Individual and Family Grants | $3.7 million as of 2/4/02 |
| Disaster Unemployment Insurance | $4.8 million to 2,350 workers as of 2/4/02 |
| Food Stamps | $3.8 million to 33,000 individuals as of 2/4/02 |
| Crisis Counseling | $154.7 million in two grants as of 8/13/02 |
| Emergency Assistance for Urban Search and Rescues and mission assignments for other federal agencies | $245 million as of 2/4/02 |
| Debris Removal | $2.5 billion as of 2/4/02 |
| Grants to the NYC Sanitation Department for Heavy Equipment for Cleanup of World Trade Center site | $66.7 million |
| NYC Office Chief Medical Examiner for ongoing forensic analysis | $10 million as of 7/15/02 |
| NYC Board of Education to clean and restore three public high schools | $4 million |
| NYC Department of Environmental Protection to monitor air quality at World Trade Center site and the city's five boroughs | $1.7 million |

*Source:* FEMA, www.fema.gov

The NYPD report did not pass judgment on the success or failure of the NYPD on September 11 but rather assessed the NYPD's response objectives and instruments in order to identify 20 improvement opportunities for the NYPD, of which six merited immediate action:

- Clearer delineation of roles and responsibilities of NYPD leaders
- Better clarity in the chain of command
- Radio communications protocols and procedures that optimize information flow
- More effective mobilization of members of the service
- More efficient provisioning and distribution of emergency and donated equipment
- A comprehensive disaster response plan, with a significant counterterrorism component (McKinsey & Company, 2002)

The Arlington County After-Action Report did declare the response by the county and others to the Pentagon terrorist attack a success that "can be attributed to the efforts of ordinary men and women performing in extraordinary fashion" (Titan Systems Corporation, 2002). The terrorist attack on the Pentagon sorely tested the plans and skills of responders from Arlington County, Virginia, other jurisdictions, and the federal government. "Notable Facts About Sept. 11 at the Pentagon" compiled in the report are provided as follows.

## NOTABLE FACTS ABOUT SEPT. 11 AT THE PENTAGON

- The first Arlington County emergency response unit arrived at the crash site less than three minutes after impact.
- Lieutenant Robert Medarios was the first Arlington County Police Department command-level official on-site. He made a verbal agreement with a representative of the Defense Protective Service that Arlington County would lead the rescue efforts of all local and federal agencies.
- More than 30 urban search-and-rescue teams, police departments, fire departments, and federal agencies assisted Arlington's police and fire in the rescue. Some of these important partners included the FBI, FEMA, the U.S. Park Police, Defense Protective Service, the Military District of Washington, the Metropolitan Washington Airport Authority, the Virginia Department of Emergency Management and USAR teams from Albuquerque, New Mexico; Fairfax County, Virginia; Montgomery County, Maryland; and Memphis, Tennessee.
- Captain Dennis Gilroy and the team on Foam Unit 161 from the Fort Meyer Fire Station were on-site at the Pentagon when Flight 77 crashed into the building. Firefighters Mark Skipper and Alan Wallace who were next to the unit received burns and lacerations but immediately began helping Pentagon employees, who were trying to escape from harm's way, out of the first floor windows.
- Captain Steve McCoy and the crew of Engine 101 were on their way to fire staff training in Crystal City when they saw the plane fly low overhead and an explosion from the vicinity of the Pentagon. McCoy was the first person to call Arlington County's Emergency Communications Center to report the plane crash.
- The Arlington County American Red Cross Chapter coordinated support from the Red Cross. The chapter had 80 trained volunteers at the time of the attack, but the organization's mutual-aid arrangements with other chapters garnered nearly 1,500 volunteers, who helped support the emergency services personnel, victims, and their families.
- Business supporters set up temporary food service on the Pentagon parking lot for rescue workers. More than 187,940 meals were served to emergency workers. Many other businesses brought phones for rescuers to call home, building materials, and other vital necessities.
- More than 112 surgeries on nine burn victims were performed in three weeks. One of the nine burn victims died after having more than 60 percent

of her body burned. There were 106 patients that reported to area hospitals with various injuries.

- 189 people died at the Pentagon: 184 victims and five terrorists.
- On the morning of September 11, 1941, the original construction on the Pentagon began.

*Source:* "Arlington County After-Action Report on the Response to the September 11 Terrorist Attack on the Pentagon," prepared for Arlington County, Virginia, by Titan Systems Corporation.

The Arlington County report contains 235 recommendations and lessons learned. Of these many recommendations, the report highlights examples of lessons learned in two categories: (1) things that worked well and contributed to the overall success of the response and (2) challenges encountered and overcome by responders that could serve as examples for other jurisdictions in the future. These lessons learned are provided.

## LESSONS LEARNED AT THE PENTAGON

The Arlington County After-Action Report contains 235 recommendations and lessons learned, each of which must be understood within the context and setting of the Pentagon response. Some specifically apply to a particular response element or activity. Others address overarching issues that apply to Arlington County and other jurisdictions, particularly those in large metropolitan areas. They have not been weighted or prioritized. This is a task best left to those with operational responsibilities and budgetary authority.

### Capabilities Others Should Emulate

1. *ICS and Unified Command.* The primary response participants understood the ICS, implemented it effectively, and complied with its provisions. The ACFD, an experienced ICS practitioner, established its command presence literally within minutes of the attack. Other supporting jurisdictions and agencies, with few exceptions, operated seamlessly within the ICS framework. For those organizations and individuals unfamiliar with the ICS and Unified Command, particularly the military, which has its own clearly defined command and control mechanisms, the Incident Commander provided explicit information and guidance early during the response and elicited their full cooperation.
2. *Mutual Aid and Outside Support.* The management and integration of mutual-aid assets and the coordination and cooperation of agencies at all government

continues

echelons, volunteer organizations, and private businesses were outstanding. Public safety organizations and chief administrative officers (CAOs) of nearby jurisdictions lent their support to Arlington County. The response to the Pentagon attack revealed the total scope and magnitude of support available throughout the Washington Metropolitan Area and across the nation.

3. *Arlington County CEMP*. The CEMP proved to be what its title implies. It was well thought out, properly maintained, frequently practiced, and effectively implemented. Government leaders were able to quickly marshal the substantial resources of Arlington County in support of the first responders, without interfering with tactical operations. County Board members worked with counterparts in neighboring jurisdictions and elected federal and state officials to ensure a rapid economic recovery, and they engaged in frequent dialogue with the citizens of Arlington County.
4. *Employee Assistance Program*. At the time of the Pentagon attack, Arlington County already had in place an aggressive, well-established EAP offering critical incident stress management (CISM) services to public safety and other county employees. In particular, the ACFD embraced the concept and encouraged all of its members to use EAP services. Thus, it is not surprising that the EAP staff was well received when they arrived at the incident site within three hours of the attack. During the incident response and in follow-up sessions weeks afterward, the EAP proved invaluable to first responders, their families, and the entire county support network. This is a valuable resource that must be incorporated in response plans.
5. *Training, Exercises, and Shared Experiences*. The ACED has long recognized the possibility of a weapons of mass destruction (WMD) terrorist attack in the Washington Metropolitan Area and has pursued an aggressive preparedness program for such an event, including its pioneering work associated with the MMRS. In preparation for anticipated problems associated with the arrival of Y2K, Arlington County government thoroughly exercised the CEMP. In 1998, the FBI Washington Field Office (WFO) established a fire liaison position to work specifically with area fire departments. Washington Metropolitan Area public safety organizations routinely work together on events of national prominence and shared jurisdictional interests, such as presidential inaugural celebrations, Heads of State visits, international conferences such as the periodic International Monetary Fund (IMF) conference, and others. They also regularly participate in frequent training exercises, including those hosted by the Pentagon and MDW. All this and more contributed to the successful Pentagon response.

## Challenges That Must Be Met

1. *Self-Dispatching*. Organizations, response units, and individuals proceeding on their own initiative directly to an incident site, without the knowledge and

permission of the host jurisdiction and the Incident Commander, complicate the exercise of command, increase the risks faced by bonafide responders, and exacerbate the challenge of accountability. WMD terrorist event response plans should designate preselected and well-marked staging areas. Dispatch instructions should be clear. Law enforcement agencies should be familiar with deployment plans and quickly establish incident site access controls. When identified, self-dispatched resources should be immediately released from the scene, unless incorporated into the Incident Commander's response plan.

2. *Fixed and Mobile Command and Control Facilities.* Arlington County does not have a facility specifically designed and equipped to support the emergency management functions specified in the CEMP. The conference room currently used as the EOC does not have adequate space and is not configured or properly equipped for that role. The notification and recall capabilities of the Emergency Communications Center are constrained by equipment limitations, and there are no protected telephone lines for outside calls when the 911 emergency lines are saturated. The ACED does not have a mobile command vehicle and relied on the use of vehicles belonging to other organizations and jurisdictions. The ACPD mobile command unit needs to be replaced or extensively modernized.
3. *Communications.* Almost all aspects of communications continue to be problematic, from initial notification to tactical operations. Cellular telephones were of little value in the first few hours and cellular priority access service (CPAS) is not provided to emergency responders. Radio channels were initially oversaturated, and interoperability problems among jurisdictions and agencies persist. Even portable radios that are otherwise compatible were sometimes preprogrammed in a fashion that precluded interoperability. Pagers seemed to be the most reliable means of notification when available and used, but most firefighters are not issued pagers. The Arlington County EOC does not have an installed radio capacity and relied on portable radios coincidentally assigned to staff members assigned duties at the EOC.
4. *Logistics.* Arlington County, like most other jurisdictions, was not logistically prepared for an operation of the duration and magnitude of the Pentagon attack. The ACED did not have an established logistics function, a centralized supply system, or experience in long-term logistics support. Stock levels of personal protective equipment (PPE), critical high-demand items (such as batteries and breathing apparatus), equipment for reserve vehicles, and medical supplies for EMS units were insufficient for sustained operations. These challenges were overcome at the Pentagon with the aid of the more experienced Fairfax County Fire and Rescue Department logistics staff. A stronger standing capacity, however, is needed for a jurisdiction the size of Arlington County.

*continues*

5. *Hospital Coordination.* Communications and coordination were deficient between EMS control at the incident site and area hospitals receiving injured victims. The coordination difficulties were not simple equipment failures. They represent flaws in the system present on September 11. Regional hospital disaster plans no longer require a Clearinghouse Hospital or other designated communications focal point for the dissemination of patient disposition and treatment information. Thus, hospitals first learned of en route victims when they were contacted by transporting EMS units, and EMS control reconstructed much of the disposition information by contacting hospitals after the fact. Although the number of victims of the Pentagon attack were fewer than many anticipated, they were not insignificant. An incident with more casualties would have seriously strained the system.

*Source:* "Arlington County After-Action Report on the Response to the September 11 Terrorist Attack on the Pentagon," prepared for Arlington County, Virginia, by Titan Systems Corporation.

The events at the World Trade Center and the Pentagon vary significantly in size and impact but, from a responder's perspective, they are similar in terms of surprise and challenges. There are striking similarities between the improvement opportunities listed in the NYPD report and the lessons learned in the Arlington County report. While the specifics vary, both responses identified issues in five key areas:

- Command
- Communications
- Coordination
- Planning
- Dispatching personnel

Many of the actions taken after September 11 by government officials and emergency managers at the federal, state, and local levels reflect the need for changes in order to prepare for the next terrorist event.

# FEDERAL GOVERNMENT TERRORISM ACTIVITY

For FEMA and its partner agencies in the Federal Response Plan (FRP), the most significant action taken by the federal government to combat terrorism is the proposed creation of the Department of Homeland Security (DHS).

For state and local emergency managers, the most significant result of federal government actions since September 11 has been the increased funding and additional funding agencies providing support for first responders and emergency management terrorism planning and prevention activities.

For the American people, the most significant effect of federal government activities to combat terrorism is the confusion resulting from the terrorism threat warnings being issued by public officials. All three perspectives are discussed in this section.

## The Department of Homeland Security

On November 25, 2002, President Bush signed into law the Homeland Security Act of 2002 (Public Law 107-296; referred to as 'HS Act' herein), and announced that former Pennsylvania Governor Tom Ridge would be Secretary of a new Department of Homeland Security (DHS) to be created. This act, which authorizes the greatest federal government reorganization since President Harry Truman, joined the various branches of the armed forces under the Department of Defense, is charged with a three-fold mission of protecting the United States from further terrorist attacks, reducing the nation's vulnerability to terrorism, and minimizing the damage from potential terrorist attacks and natural disasters.

The creation of DHS is the culmination of an evolutionary legislative process that began largely in response to criticism that increased federal intelligence inter-agency cooperation could have prevented the September 11th terrorist attacks. A sweeping reorganization into the new Department, which officially opened its doors on January 24, 2003, will join together over 179,000 federal employees from twenty-two existing federal agencies under the umbrella of a single, Cabinet-level organization.

Beginning March 1, 2003, almost all of the federal agencies (and their respective employees) named in the act are expected to begin their move, whether literally or symbolically, into the new Department. Those remaining will follow on June 1, 2003, with all incidental transfers completed by September 1, 2003.

The DHS organizational structure is presented in Figure 9-1. A total of 22 federal agencies and programs will be transferred into the DHS including the U.S. Customs Service (Department of Treasury), Immigration and Naturalization Service and Border Patrol (Department of Justice), Animal and Plant Health Inspection Service (Department of Agriculture), Transportation Security Administration (Department of Transportation), Federal Protective Service (General Services Administration), FEMA, and the U.S. Coast Guard.

While a handful of these agencies will remain intact after the move, most will be incorporated into one of four new directorates; Border and Transportation Security (BTS), Information Analysis and Infrastructure Protection (IAIP), Emergency Preparedness and Response (EP&R), and Science and Technology (S&T). A fifth directorate, Management, will not incorporate any existing federal agencies.

Several existing agencies that are being transferred into the Department of Homeland Security intact, and others that are to be newly created, will report directly to the Office of the Secretary. Most notable of these agencies are the U.S. Coast Guard, the U.S. Secret Service, the Office of State and Local Government Coordination, the Office of the Inspector General, and the Office of the Deputy Secretary.

Office of State and Local Government Coordination, once established, will work to ensure that close coordination takes place with state and local first responders, emergency services, and governments in all DHS programs and

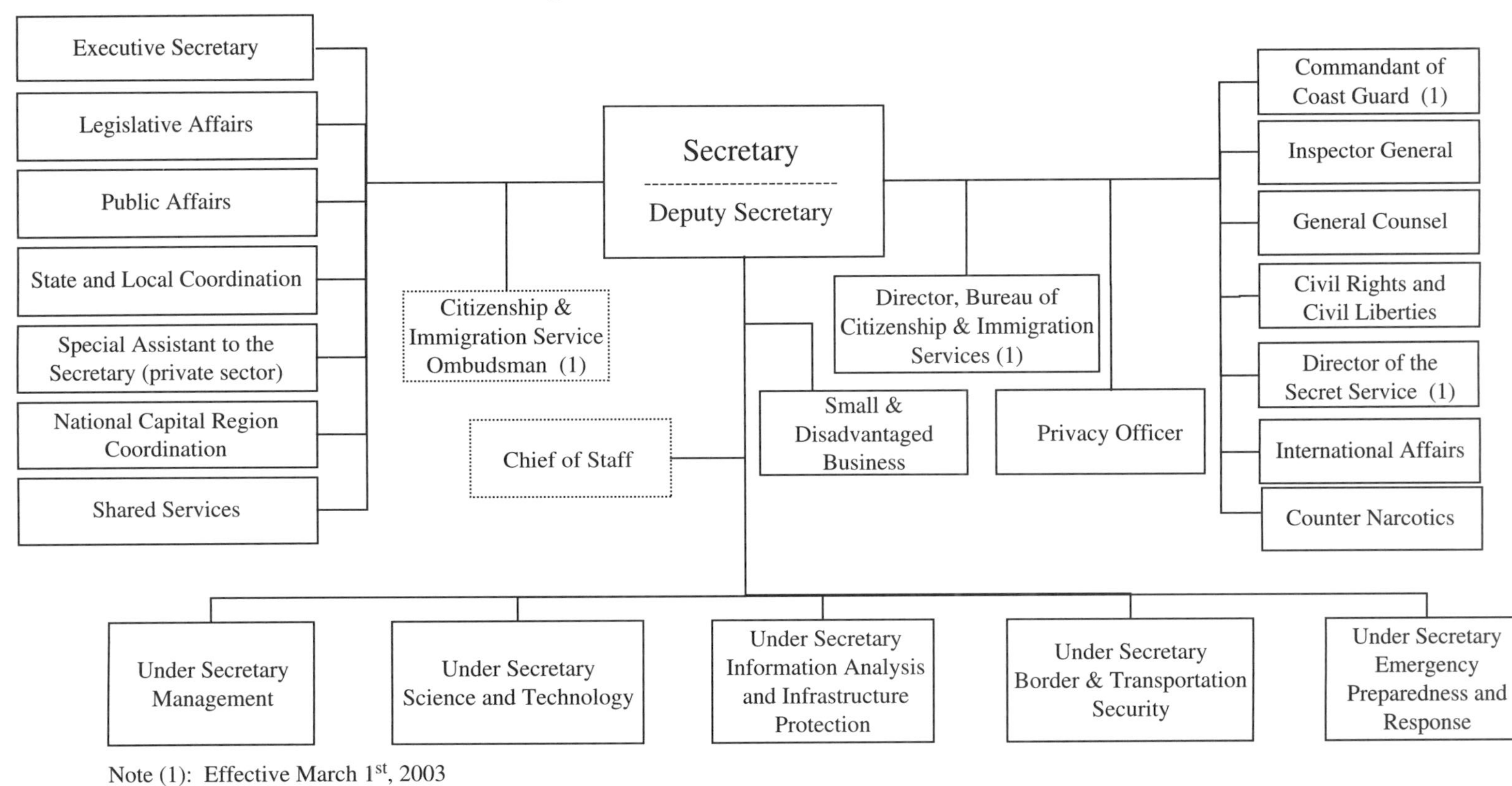


**Figure 9-1** Department of Homeland Security Organizational Chart.

activities. It is important to note that the primary responsibility for investigating and prosecuting acts of terrorism remains in Federal, State, and local law enforcement agencies except as specifically transferred to DHS by the Homeland Security Act of 2002.

The Emergency Preparedness and Response (EP&R) Directorate will be led by Under Secretary Michael Brown, former Deputy Director and General Counsel of the Federal Emergency Management Agency (FEMA), after FEMA is incorporated into EP&R on March 1, 2003. EP&R will assume FEMA's responsibilities, namely preparing for and responding to natural and technological disasters and terrorist attacks. Like FEMA, EP&R will coordinate with local and State first responders to manage disasters requiring federal government assistance, and to recover from their damaging effects. The Directorate will continue to practice a comprehensive, risk-based approach, employing a program of preparedness, prevention (mitigation), response, and recovery.

EP&R hopes to continue FEMA's mission of 'proactively help communities and citizens avoid becoming victims' (rather than relying wholly upon response), utilizing public education efforts and volunteerism to achieve this goal. Federal grants to promote the 'disaster-resistant communities' concept, partnering public and private sectors to reduce risk in disaster prone areas, will be continued. EP&R will also continue to develop a curriculum and manage the training and evaluation of local, State, and federal emergency responders.

It is anticipated that EP&R will no longer be using the Federal Response Plan (FRP) as a framework for federal government disaster response. Under the FRP, which is invoked when the President declares a national disaster, various agencies throughout the government are charged with mission-specific tasks related to response and recovery. FEMA was tasked by the FRP with, among other responsibilities, coordinating the response of all federal agencies. The FRP was used successfully after the Oklahoma City bombing (see Figure 9-2).

While this coordination responsibility will likely remain in the EP&R Directorate after the FRP is replaced with the new 'Incident Response Plan,' there are likely to be many changes to the way in which non-DHS agencies become involved. It is important to note that while EP&R will no longer be responsible for the training of first responders for terrorism-related events (whether conventional, biological, chemical, or nuclear), it will be responsible for coordinating the federal government response to them. From a response perspective, in reviewing the proposed responsibilities of the Emergency Preparedness and Response Division, note the references to the five common issues found in the reports on the NYPD performance on September 11 and the Arlington County response to the Pentagon attack. The DHS document includes reference to command in assigning itself responsibility for the command element of any federal government response to provide support for state and local efforts.

EP&R will continue to offer three mitigation grant programs currently managed by FEMA: the Hazards Mitigation Grant Program, the Pre-Disaster Mitigation Program, and the Flood Mitigation Assistance Program, and the US Fire Administration Grants, though it is expected that several procedural changes to these grants may occur.

There are two other elements within DHS that will impact emergency management, particularly at the State and local level. These are the Office of State and Local

**Figure 9-2** Oklahoma City, OK, April 26, 1995—Scene of the devastation following the Oklahoma City bombing. FEMA News Photo.

Government Coordination and the Office of Domestic Preparedness, within the Directorate for Border and Transportation Security.

The Office of State and Local Government Coordination, once established, will work to ensure that close coordination takes place with state and local first responders, emergency services, and governments in all DHS programs and activities. It is important to note that the primary responsibility for investigating and prosecuting acts of terrorism remains in Federal, State, and local law enforcement agencies except as specifically transferred to DHS by the Homeland Security Act of 2002.

The Office of Domestic Preparedness will coordinate preparedness efforts at the Federal level and work with State and local emergency response providers on all matters pertaining to terrorism including training, exercises and equipment support. This Office will provide all of the grant funding to support State and local risk

**Figure 9-3** New York, NY, September 27, 2001—The remaining section of the World Trade Center is surrounded by a mountain of rubble following the September 11 terrorist attacks. Photo by Bri Rodriguez/ FEMA News Photo.

analysis and it is tasked to cooperate closely with EP&R to prepare for and mitigate the effects of non-terrorist related disasters in the U.S.

## Funding for First Responders and Emergency Management

For state and local government, the events of September 11 (see Figure 9-3) resulted in an extraordinary increase in funding for first responders—fire, police, and emergency medical technicians—and emergency management activities. Also, the number of federal government agencies and programs now providing funds for these activities has increased significantly. In the first-responder community, only the police have historically received significant funding from the federal government. Fire departments across the country have traditionally raised most of their funding from local sources. Emergency medical technicians are often private contractors paid for by local and state government sources.

Proper training and equipping of firefighters responding to a biochemical terrorist attack has been a concern among the fire services community and FEMA since the early 1990s. Passage of the Fire Prevention and Assistance Act in 2000 was the first effort by Congress to support the nation's paid and volunteer fire departments. In Spring 2001, FEMA initiated a new Fire Grant program that provided $100 million in small grants to local fire departments for equipment, protective gear, training, and prevention programs. In 2002, the amount available for FEMA fire grants increased to $300 million. In addition to the annual fire grants, the bulk of the $3.5 billion currently proposed in FEMA's fiscal year 2003 budget has been designated for

**Figure 9-4** New York, NY, October 4, 2001—NY Firefighter chief at the site of the World Trade Center. Photo by Andrea Booher/ FEMA News Photo.

equipping and training first responders for future terrorist events (see Figure 9-4). However, upon passage of the 2003 appropriations legislation, the funds to support State and local public safety and emergency management activities were reduced.

FEMA is not the only source of terrorism funding for state and local governments. The DOJ, through a variety of programs, is making funding available for the acquisition of equipment and technology. The DHHS is making substantial funding available to state and local governments to address the threat of biochemical terrorist attacks. The CDC is providing funding for public health planning and capacity building and bolstering the national pharmaceutical stockpile. The DOD currently provides funding for emergency management training for military personnel and community officials. A partial list of federal funding sources for state and local governments in the fight against terrorism is presented.

## Communicating Threat Information to the American People

As noted earlier, a stated priority of the White House's proposed DHS is to "be the organization that coordinates provision of specific threat information to local law enforcement and sets the national threat level" (White House Office of Homeland Security, 2002). To date, this task has fallen on the Office of Homeland Security in the White House and the DOJ. Initial threat warnings were found to be confusing by the American people, and a new system was developed and deployed in March 2002.

## HOMELAND SECURITY ADVISORY SYSTEM

As part of a series of initiatives to improve coordination and communication among all levels of government and the American public in the fight against terrorism, President Bush signed Homeland Security Presidential Directive 3, creating the Homeland Security Advisory System (HSAS). The advisory system will be the foundation for building a comprehensive and effective communications structure for disseminating information regarding the risk of terrorist attacks to all levels of government and the American people.

The HSAS establishes five Threat Conditions with associated suggested Protective Measures:

### Low Condition—Green

Low risk of terrorist attacks. The following Protective Measures may be applied:

- Refining and exercising preplanned Protective Measures
- Ensuring personnel receive training on HSAS, departmental, or agency-specific Protective Measures
- Regularly assessing facilities for vulnerabilities and taking measures to reduce them

### Guarded Condition—Blue

General risk of terrorist attack. In addition to the previously outlined Protective Measures, the following may be applied:

- Checking communications with designated emergency response or command locations
- Reviewing and updating emergency response procedures
- Providing the public with necessary information

### Elevated Condition—Yellow

Significant risk of terrorist attacks. In addition to the previously outlined Protective Measures, the following may be applied:

- Increasing surveillance of critical locations
- Coordinating emergency plans with nearby jurisdictions

*continues*

- Assessing further refinement of Protective Measures within the context of the current threat information
- Implementing, as appropriate, contingency and emergency response plans

### High Condition—Orange

High risk of terrorist attacks. In addition to the previously outlined Protective Measures, the following may be applied:

- Coordinating necessary security efforts with armed forces or law enforcement agencies
- Taking additional precaution at public events
- Preparing to work at an alternate site or with a dispersed workforce
- Restricting access to essential personnel only

### Severe Condition—Red

Severe risk of terrorist attacks. In addition to the previously outlined Protective Measures, the following may be applied:

- Assigning emergency response personnel and prepositioning specially trained teams
- Monitoring, redirecting, or constraining transportation systems
- Closing public and government facilities
- Increasing or redirecting personnel to address critical emergency needs

*Source:* White House Website

## STATE GOVERNMENT TERRORISM ACTIVITY

A good way to understand state government activities in terrorism is to examine the priorities set by the nation's governors and state emergency managers. A survey of State Homeland Security Structures by the National Emergency Management Agency (NEMA) conducted in June 2002 found that "all 50 states maintain a primary point of contact for antiterrorism/homeland security efforts":

- Governor's office—12 states
- Military/Adjutant General—12 states
- Emergency Management—10 states
- Public Safety—9 states
- Law Enforcement—3 states
- Attorney General—2 states
- Lt. Governor—2 states
- Land Commissioner—1 state (National Emergency Management Association 2002)

**Figure 9-5** New York, NY, November 1, 2001—FEMA's Disaster Field Office in New York has been ground zero for the agency's operations in the aftermath of the World Trade Center tragedy. Photo by Larry Lerner/ FEMA News Photo.

In August 2002, the NGA Center for Best Practices of the National Governors Association released "States' Homeland Security Priorities." A list of 10 "major priorities and issues" was identified by the NGA Center through a survey of states' and territories' state homeland security offices (NGA Center for Best Practices, 2002). A list of these priorities is presented as follows.

## LIST OF STATES' HOMELAND SECURITY PRIORITIES

- Coordination must involve all levels of government.
- The federal government must disseminate timely intelligence information to the states.
- States must work with local governments to develop interoperable communications between first responders, and adequate wireless spectrum must be set aside to do the job.
- State and local governments need help and technical assistance to identify and protect critical infrastructure.
- Both the states and federal government must focus on enhancing bioterrorism preparedness and rebuilding the nation's public health system to address 21st-century threats.

*continues*

- The federal government should provide adequate federal funding and support to ensure that homeland security needs are met.
- The federal government should work with states to protect sensitive security information, including restricting access to information available through "freedom of information" requests.
- An effective system must be developed that secures points of entry at borders, airports, and seaports without placing an undue burden on commerce.
- The National Guard has proven itself to be an effective force during emergencies and crises. The mission of the National Guard should remain flexible, and Guard units should primarily remain under the control of the governor during times of crises.
- Federal agencies should integrate their command systems into existing state and local incident command systems (ICS) rather than requiring state and local agencies to adapt to federal command systems.

*Source:* NGA Center for Best Practices, Issue Brief, August 19, 2002

Coordination, command, information sharing, funding support, and interoperability of communications are some of the critical issues identified by the NGA. Use of the National Guard, resource dispatch issues, support for bioterrorism preparedness, and the rebuilding of the public health system are also critical issues.

On October 1, 2001, NEMA released a "White Paper on Domestic Preparedness" that was supported by the Adjutants Generals Association of the United States, the International Association of Emergency Managers (which represents local emergency management officials), and the National Guard Association of America. In this White Paper, it was stated that, "NEMA thinks it critical that the following enhancements be incorporated into a nationwide strategy for catastrophic disaster preparedness" (National Emergency Management Association, 2001). A total of 22 enhancements were presented in the White Paper in three general categories: emergency preparedness and response, health and medical, and additional WMD (Weapons of Mass Destruction) recommendations. A partial list of these enhancements is provided.

## PARTIAL LIST OF ENHANCEMENTS PRESENTED IN THE NEMA PAPER ON DOMESTIC PREPAREDNESS

### Emergency Preparedness and Response

- Congress should provide to the states immediate federal funding for full-time catastrophic disaster coordinators in moderate and high-risk local jurisdictions of the United States.
- States need financial assistance to improve catastrophic response and Continuity of Operations Plans (COOP) and Continuity of Government (COG) for states.

- Interstate and intrastate mutual assistance must be recognized and supported by the federal government as an expedient, cost-effective approach to disaster response and recovery.
- FEMA, state, and local emergency managers must implement renewed emphasis on family and community preparedness to ensure that Americans have necessary skills to survive a catastrophic disaster.
- A standardized national donations management protocol is needed to address the outpouring of food, clothing, supplies, and other items that are commonly sent to affected state localities following a disaster.

## Health and Medical

- The medical surge capacity must be strengthened. The emergency management, medical, and public health professions must work with lawmakers to ensure that each region of the United States has a certain minimum surge capacity to deal with mass casualty events.
- State-Local Disaster Medical Assistance Teams should be developed across the country with standardized equipment, personnel, and training.

## Additional WMD Recommendations

- The DOJ should immediately release the FY00 and FY01 equipment funds in order to begin implementing these recommendations, and then require a basic statewide strategy in order to receive FY02 funds; and further, provide funding to states to administer the equipment program.
- Congress and the DOD should authorize homeland defense as a key federal defense mission tasking for the National Guard.
- State-Local Urban Search and Rescue capabilities should be developed across the country with the standardized equipment, personnel, and training.
- National interagency and intergovernmental information management protocols are needed to support information sharing (i.e., Damage/Situation Reports, Warning/Intelligence Reports, Resource Coordination).
- Better federal interagency coordination is needed to assist states in identifying and accessing the full range of federal resources and assistance available to them.
- FEMA's fire grant program should be expanded and modified to strengthen regional and national, not just local, fire protection capabilities to respond to catastrophic disasters.
- There is a need for technology transfer from the federal government and technology contractors to state and local governments to support an automated decision support system.

*Source:* National Emergency Management Association, White Paper of Domestic Preparedness, October 1, 2001

The NEMA White Paper presents "enhancements" that address coordination, communications, command, and information sharing, funding, technology, and public health system and preparedness issues. Also included were the use of National Guard assets and increasing the capabilities of State-Local Urban Search and Rescue. Expanding the FEMA fire grant program and establishment of standardized national donations protocol was included in the White Paper.

Many states are moving ahead in terrorism and homeland security planning and other activities. A report compiled by the White House Office of Homeland Security found activities in many states, cities, and counties in the following four general areas (White House Office of Homeland Security, 2002):

- Developing plans
- Information sharing
- Responding to biological threats
- Protecting critical infrastructure

## LOCAL GOVERNMENT TERRORISM ACTIVITY

Emergency preparedness, mitigation, response, and recovery all occur at the local community level. This is true for terrorism preparedness, mitigation, response, and recovery activities as well. It is at the local level that the critical planning, communications, technology, coordination, command, and spending decisions matter the most. The priorities of groups such as the National Conference of Mayors and the National Associations of Counties (NACo) represent what matters at the local community level in the fight against terrorism. The fight against terrorism has spawned a series of new requirements in preparedness and mitigation planning at the local level.

The NACo has created a "Policy Agenda to Secure the People of America's Counties." This policy paper states: "Counties are the first responders to terrorist attacks, natural disasters and major emergencies" (NACo, 2001). The NACo has established a 43-member NACo Homeland Security Task Force that on October 23, 2001 prepared a set of 20 recommendations in four general categories concerning homeland security issues. The four general areas are public health, local law enforcement and intelligence, infrastructure security, and emergency planning and public safety. The titles of each of the 20 NACo recommendations are presented as follows.

### RECOMMENDATIONS FOR COUNTIES AND HOMELAND SECURITY PREPARED BY NACo HOMELAND SECURITY TASK FORCE

#### Public Health

1. Fund the Public Health Threats and Emergencies Act.
2. Improve the Health Alert Network.

3. Ensure an adequate supply of vaccines and antibiotics.
4. Develop a national policy to prioritize medical treatment.
5. Train health personnel.
6. Ensure that adequate medical surge capacity exists.

### Local Law Enforcement and Intelligence

7. Authorize a local antiterrorism block grant.
8. Include counties in antiterrorism task forces.
9. Balance heightened border security with economic activity.

### Infrastructure Security

10. Reimburse counties for airport security costs.
11. Assist ports and transit systems in financing security measures.
12. Help localities secure public utilities and a safe water supply.
13. Include security in infrastructure development.
14. Reimburse counties for costs incurred on behalf of the federal government.
15. Assist counties to develop evacuation capacity.

### Emergency Planning and Public Safety

16. Train county officials to prepare for and respond to acts of terror.
17. Assist public safety communications interoperability and interference issues.
18. Establish a Public Communication Network.
19. Urge the release of federal research to assist counties.
20. Provide immunity to encourage mutual aid and support.

*Source:* "Policy Agenda to Secure the People of America's Counties," National Association of Counties, October 2001

In December 2001, the U.S. Conference of Mayors released "A National Action Plan for Safety and Security in America's Cities." The document was prepared as part of the Mayors Emergency Safety and Security Summit held in Washington, D.C., on October 23–25, 2001. It contains recommendations in four priority areas: transportation security, emergency preparedness, federal-local law enforcement, and economic security. In this document, the mayors made the following critical point:

> It is important to understand that while the fourth area, economic security, is viewed as the ultimate goal of a nation, it cannot be achieved in the absence of the first three. That

> is, securing our transportation system, maximizing our emergency response capability, and coordinating our law enforcement response to threats and incidents at all levels are viewed as prerequisites to eliminating the anxiety that has accelerated the nation's economic downturn, and to achieving economic security for the nation. ("A National Action Plan for Safety and Security in America's Cities," The U.S. Conference of Mayors, December 2001)

In the section on emergency preparedness, the mayors' plan presents recommendations for action in the following areas:

- Office of Homeland Security
- Reimbursement for heightened security
- Metropolitan emergency management
- Communications technology
- Protective equipment/training—Direct local assistance
- Public health system
- Coordination
- Communications
- Training
- Personnel
- Facilities
- Equipment/supplies
- Stadium/arena security
- Water and wastewater security

The principal areas of concern in federal-local law enforcement for the mayors are communications, coordination, and border city security. In the transportation security section, the mayors' paper presents recommendations concerning security issues in each of the major transportation modes: airport, transit, highway, rail, and port.

Both the NACo and the mayors' policy papers identify issues in the areas of command, coordination, communications, funding, and equipment, training, and mutual aid. These two papers also raise concerns about critical community infrastructure, including the public health system, which is maintained and secured at the local level of government.

The events of September 11 established the security of community infrastructure as a potential target for terrorist attacks. Community infrastructure has always been vulnerable to natural and other technological disaster events, so much so that FEMA's largest disaster assistance program, Public Assistance, is designed to fund the rebuilding of community infrastructure damaged by a disaster event. Local government officials and local emergency managers must now increase the attention they give to protecting and securing community infrastructure from a terrorist attack. They must also include in these preparedness efforts the local public health system. A checklist designed for the City of Boone (NC) as part of a Technological Annex developed for the town's All-Hazards Planning and Operations Manual in March 2002 is provided.

## GOALS AND QUESTIONS FOR LOCAL GOVERNMENTS PREPARING FOR TERRORISM—BOONE, NORTH CAROLINA

The preparedness and response of terrorist events requires that local governments do the following:

- Identify the types of events that might occur in the community.
- Plan emergency activities in advance to ensure a coordinated response.
- Build the capabilities necessary to respond effectively to the consequences of terrorism.
- Identify the type or nature of an event when it does happen.
- Implement the planned response quickly and efficiently.
- Recover from the incident.

The response to terrorism is similar in many ways to that of other natural or man-made disasters for which Boone has already prepared. With additions and modifications, the development of a completely separate system can be avoided. Training and public education are vital, and understanding the federal assistance available will drastically increase local capacity before and during a terrorist attack.

The following are the general types of activities that Boone must undertake to meet the aforementioned objectives:

- Strengthen information and communications technology.
- Establish a well-defined incident command structure that includes the FBI.
- Strengthen local working relationships and communications.
- Educate the healthcare and emergency response community about identification of bioterrorist attacks and agents.
- Educate the healthcare and emergency response community about medical treatment and prophylaxis for possible biological agents.
- Educate the local health department about state and federal requirements and assistance.
- Maintain a locally accessible supply of medications, vaccines, and supplies.
- Address healthcare worker safety issues.
- Designate a spokesperson to maintain contact with the public.
- Develop comprehensive evacuation plans.
- Become familiar with state and local laws relating to isolation/quarantine.
- Develop or enhance local capability to prosecute crimes involving weapons of mass destruction or the planning of terrorism events.
- Develop, maintain, and practice an infectious diseases emergency response plan.
- Practice with surrounding jurisdictions/strengthen mutual agreement plans.

*continues*

- Outline the roles of federal agency assistance in planning and response.
- Educate the public in recognizing events and how to respond as individuals.
- Stay current.

*Source:* Town of Boone All-Hazards Planning and Operations Manual, Technological Hazards Annex, March 2002

Local officials must also understand how to work with federal law enforcement officials if a terrorist incident occurs in the community. Information from the Boone Technological Hazards Annex on the process of transitioning from fire to law enforcement to FBI as part of the incident command system is provided.

## INCIDENT COMMAND SYSTEM STRUCTURE: PROCESS OF TRANSITIONING FROM FIRE TO LAW ENFORCEMENT TO FBI

Local FBI should be the first contact following an attack. That call will lead to the DOD, DOJ, DHHS, DOE, FEMA, EPA, and many others. The FBI will set up a Strategic Information Operations Center (SIOC) to provide constant direction in the attack response. If other agencies are involved, then the duties of the SIOC will be expanded into a Joint Operations Center (JOC). This will address the areas of command, operations, support, and consequence management, and will include a Joint Information Center (JIC) to provide information to the media and the public.

FEMA will contact the governor and the President to determine if federal assistance is necessary. They will establish a Regional Operations Center (ROC) for a deployed Emergency Response Team (EST).

The local government is responsible for the first consequence management. This includes measures to protect public health and safety, restore essential government services, and provide emergency relief to governments, businesses, and individuals affected by the consequences of terrorism. The logical steps in coordination and command that should be followed are detailed as follows:

- When fire is in command, during the initial response, law enforcement (LE) follows the principles for assisting with a mass casualty incident.
- When command passes to LE, the officer in charge will set the goals for the operation and begin the preliminary criminal investigation.
- When fire begins to wind down, the fire and LE commanders will agree when the incident command passes to LE. When this occurs, simultaneous broadcasts should be made on both fire and LE channels so that all personnel understand the pass of command.
- The specific location for the LE command post should be repeated via radio so there is no mistaking the location.

- It is essential that a command level fire official remain in the command post to ensure continuity of information and to provide an officer who can direct fire resources if they are needed during this time.
- It is important to remember that, until the call is made that the incident is an act of domestic or foreign terrorism, the event is the local jurisdiction's homicide, assault, vandalism, bombing, etc.

The FBI will usually dispatch an initial special agent as soon as the incident occurs, as a direct result of just being in your town and monitoring radio frequencies. If there is even a slight suspicion of terrorism, the FBI needs to be contacted. The time for the special agent to arrive is from minutes to several hours. It is better to have the FBI on scene as soon as possible because they speed up the investigation if it turns out to be actual terrorism and simply act as another highly trained official if it is just a homicide. The FBI also provides technical support and resources not normally found at the local level.

When the FBI team is assembled and in place, it assumes command. The FBI has its own policies about evidence collection, so if terrorism is suspected, it is best to seal the area and let the FBI evidence technicians collect the evidence when they arrive, to avoid a later chain-of-custody issue of evidence. If some evidence needs to be preserved immediately, this should be done for later transfer to FBI custody.

The FBI will still need continued support and assistance from local LE and Fire. They need help with traffic control, body and evidence recovery, scene security, and a host of other critical tasks. They will create a JOC to coordinate the federal response to the event, similar to the EOC. All federal agencies will report to and act out of this office. If possible, they will try to co-locate the JOC and EOC. At a minimum, they will try to provide a command-level officer to stay in the local agency EOC to facilitate direct communication between the groups involved in the Unified Command and to facilitate sharing of information resources and personnel.

It is best to work out of one central EOC. The FBI's JOC is based on the large number of agencies and support staff that may be called in. If the event is a large one, the Incident Commander should be looking for an area large enough to accommodate these priorities so that both facilities can be at the same location.

As the local agencies' activities wind down, fewer staff will be necessary, until the JOC will become the only coordination point for the incident. When on-scene activity ceases, all follow-up investigative activity will be the responsibility of the FBI.

This is the triangle issue, of FBI, LE, and FIRE, and during each of the phases, the three move around the triangle, sharing concerns and information in the Unified Command concept.

*Source:* Town of Boone All-Hazards Planning and Operations Manual, Technological Hazards Annex, March 2002

# CONCLUSION

Emergency management in the United States was changed forever by the events of September 11. New focus, new funding, new partners, and new concerns associated with the fight against terrorism are changing the way emergency management functions in this country every day. At the federal government level, a new Department of Homeland Security is being established that will include FEMA and all of the federal government disaster management programs. How traditional disaster response recovery and mitigation programs will fare in this new structure is uncertain. At the state level, governors and state emergency management directors are calling for better coordination, new communications technologies, and always more and more funding. At the local government level, terrorism is a new threat that greatly expands their facility security requirements and is added to a long list of needs and priorities, but the threat of terrorism can't be ignored. Issues of coordination, communications, and funding concern local governments as well.

The United States has taken its typical response to a new problem. It has reorganized and committed huge amounts of funding to reducing the problem. The ability of the DHS to achieve an enhanced level of coordination, communication, and readiness remains uncertain. Preventing future terrorism attacks is outside of the purview of the DHS and resides with the intelligence community, the military, diplomatic corps, and law enforcement. What the DHS can offer is a better-prepared and better-equipped first-responder cadre, enhanced transportation and border security, and more money for emergency management programs.

But how long will it take for the DHS to operate as a cohesive organization? In the rush to demonstrate action, both the Bush administration and Congress have failed to look carefully at the longer-term implications of these decisions. The probabilities of a massive natural disaster are equally as great as another terrorism event. Will the DHS's focus on terrorism erode progress in natural hazards risk reduction, and will the nation be ready for the next major hurricane or earthquake? An American leader once said words to the effect that if people fail to learn from their mistakes, then they are destined to repeat them. Let's hope history doesn't repeat itself.

## CASE STUDY: TOWN OF BOONE (NC), BIOTERRORISM RESPONSE TEMPLATE

Although chemical incidents would be easy to detect because of their immediate effects, biological attacks could take several days. Monitoring and first-responder training is the best defense. Boone could become a secondary location of a bioterrorism incident that occurred elsewhere if an infected person (resident or tourist) arrives without knowledge of his or her contagious infection. Several residents could be infected before it was even known that the original attack took place.

When responding to a biological attack, the timing of the response is critical. The following six elements of response must be well coordinated:

- Medical surveillance to detect the attack
- Quick, rapid, and appropriate decisions

- Implementation of preexisting response plans
- Rapid and appropriate distribution of prophylaxis
- Ability to keep up with the flow of sick and "worried well"
- Ability of response system to remove and rapidly utilize outside help

The 13 major response functions of state and local communities are detailed by the DOD Bio-weapon Improved Response Program. These functions need to be developed through training and exercise:

- Medical surveillance
- Medical diagnosis
- Epidemiological investigation
- Criminal investigation
- Mass prophylaxis
- Residual hazard assessment and mitigation
- Control affected area/population
- Care of presented casualties and "worried well"
- Fatality management
- Emergency management operations
- Resource and logistic support
- Continuity of infrastructure
- Family support services

It is essential that treatment protocols or treatment plans are designed to allow the rapid application of treatment tactics in the time-compressed and information-poor environment following an attack. EMS agencies, extrapolating from known response methodologies such as Hazmat protocols, can deal with issues of potential biological exposures as well as for hoaxes, with relatively simple adjustments.

It is crucial for information sharing and communication that each healthcare agency acquires broadband high-speed Internet access. Each administrator and point of entry for patients must have data entry terminals for rapidly entering information into its data systems. It is also important to maintain treatment plans, resources for referral on lesser-known disease entities, and decision-support systems developed locally or obtained from the CDC.

An emergency response requires instant and multifaceted communications networks that are reliable and flexible. Redundant phone systems, both analog and digital and wireless, should be put into use. This preparation step involves a comprehensive communications plan that can be rapidly expanded as the situation warrants. An ability to increase the number of dedicated wireless phones within the healthcare system should rank high on the to-do list. Broadcast fax capability can be part of the phone systems, allowing for quick dispersal of information to a wide geographic area.

The development of a network of contacts is necessary to create a sense of trust, confidence, and easy access. In order to decide whom to include, it must be determined who is needed to deal with a large-scale event such as an aerosol release of a biological agent. There should be no less than two ways to contact each individual on the list, in addition to an e-mail address. Include all individuals in at least one exercise or reaction to hoax. Examples of people who need to be included are listed as follows:

- Police and Fire
- Medical Director
- Emergency Management
- Emergency Room Managers
- Emergency Room Physicians
- Public Health Units
- Lab Managers
- Infection Control Officials at each hospital
- Pharmacy Managers
- Data Control/Information Services

Surveillance of diseases should be well known to each medical practitioner. Medical providers must stay informed regarding the latest trends in testing and treatment methods, prevention strategies, and population-based reportable diseases as they evolve. This could be included in the role of the terrorism coordinator in the emergency manager's office. Rapid confirmation tests for disease-specific entities should be acquired. This list of diseases will change over time, so it will be necessary to keep up with current releases of information from the federal government.

It is important to have on hand an inventory of locally available antibiotics and vaccines. In many incidents of bioweapon use, all persons in the affected area may need antibiotic prophylaxis, as well as specific drugs and antitoxins. Stockpiles are available anywhere in the United States within 12 hours.

The following are recommendations to specific response systems within the community:

*Hospitals*

1. Review all relevant disaster-response plans and ensure that appropriately designated staff are familiar with their content and strategies.
2. Establish internal and external lines of communication. Ensure that medical staff are aware of the need to report suspicious cases of illnesses to public health authorities, and are familiar with who these authorities are. Have in place dedicated staff, phones, and fax machines to support rapid reporting.
3. Have hospital leaders establish collaborative strategies for communicating with neighboring hospitals, civic leaders, and public health authorities.
4. Quantify pharmaceutical and antibiotic supplies, both at central and satellite facilities. Routinely update this list.
5. Assess routine staffing and emergency call-up plans and ensure that these are supported with communication and transportation strategies. Update the roster of essential personnel.
6. Maintain ongoing primary and redundant communication systems.
7. Ensure that appropriate healthcare professionals are aware of the importance of reporting unusual disease presentations, disease clusters, and atypical patterns of hospital use and know the mechanisms to do reporting.

*Physicians*

1. Develop an increased awareness of the ongoing threat of bioterrorism.
2. Become familiar with and develop a working knowledge of the most likely and dangerous pathogens as viewed by the CDC.
3. Become familiar with relevant lines of communication and important and emergency phone numbers (e.g., hospital epidemiologist, state epidemiologist, local health department), and the CDC emergency number.
4. Monitor disease patterns and patient volumes in clinics and offices. Immediately notify the appropriate authorities if you suspect an unusual event or need medical guidance.
5. Refer patients to the CDC public inquiry phone number regarding information about infectious diseases and bioterrorism preparedness response efforts.
6. Have referral numbers for mental health and support services as needed.
7. Understand the use of the National Pharmaceutical Stockpile and its large quantities of antibiotics and vaccines that could be distributed in the event of an epidemic brought on by an act of bioterrorism.

*Emergency Medical Services*

1. Ensure that leaders are generally familiar with what a bioterrorism attack might demand of civil authorities and what resources are available to meet these demands.
2. Identify and, if feasible, meet with public health and medical experts who might provide guidance to key decision makers during a public health emergency.
3. Put in place primary and backup communication systems to ensure that civil authorities can contact key medical, public health, and emergency response workers.
4. Ensure that civil authorities can quickly broadcast emergency messages, health alerts, and educational information across multiple media, including radio, television, and Websites. If older civil alert systems (e.g., air horns) are available, educate the public regarding their possible use and meaning.
5. Identify existing gaps in linkages, coordination and response, and communication among hospitals, public health agencies, and emergency response workers.
6. Develop transportation plans that facilitate movement of emergency vehicles, entrance to and exit from hospitals and care centers, and rapid deployment of essential healthcare workers from their homes or off-site locations to primary hospital and healthcare sites.
7. Designate a dedicated point of contact to receive information from medical and public health agencies in the event of a bioterrorism attack.

*Public Health Agencies*

1. Have local and state public health agencies collectively review bioterrorism response plans. Focus on ensuring the integration of response plans, including mechanisms for sharing resources and personnel as needed.

2. Establish syndrome surveillance procedures to monitor and detect atypical disease presentations and clusters. Examine both passive and active surveillance systems and refine them across public health agencies and with reporting sources.
3. Establish and maintain the capacity to accept reports of unusual disease events 24 hours a day, seven days a week. Ensure systems of continual, bidirectional communication between public health agencies and hospitals under their purview.
4. Make appropriately trained disease investigation staff available for immediate mobilization and deployment as needed. Review staffing levels and initiate plans to determine nonurgent public health functions and clinics if it is necessary to pull additional clinical and field staff for urgent investigation activities.
5. Assess communication systems, including procedures for immediately contacting public health and political leaders. Assess systems to ensure that appropriate authorities could be contacted at the outset of an emergency. Assess and test mechanisms for maintaining ongoing communication, including pagers, cellphones, and wireless e-mail systems. All staff members who provide on-call and disease investigation response and decision making should be adequately resourced for constant communication.
6. Hold regular meetings with all appropriate government and nongovernmental agencies and organizations to continually review and refine plans.
7. Utilize CDC protocol for Notification Procedures in the Event of a Bioterrorist Incident (http://www.bt.cdc.gov/EmContact/Protocols.asp)

# APPENDIX 9-1

## Contacts

**National Domestic Preparedness Office (NDPO)**
Domestic Preparedness Helpline: 800.368.6498
Main Number: 202.324.8186
Training: 202.324.0265
Exercises: 202.324.0299
Equipment: 202.324.0220
Planning: 202.324.0276
Website: www.fbi.gov/program.htm, http://php.indiana.edu/~tgatkins/bmpt.html, www.nbc-prepare.org, www.ndpo.org

**Centers for Disease Control and Prevention (CDC)**
Main Number: 800.311.3435
Disease Information: 888.232.3228
Website: www.cdc.gov

**Department of Defense (DOD)**
Main Number: 703.695.5261

Chem/Bio Helpline: 615.399.9908
Website: www.defenselink.mil

**Department of Energy (DOE)**
Main Number: 202.401.3000
Website: www.doe.gov

**Department of Justice (DOJ)**
Main Number: 202.514.2001
Community Resource Associates (CRA): 615.399.9908
National Center for Domestic Preparedness: 256.848.7043
State and local domestic preparedness support: 202.305.9887
Website: www.usdoj.gov

**Environmental Protection Agency (EPA)**
Main Number: 202.260.4700
National Response Center: 800.424.8802 (emergency)
National Response System: 202.260.8600
Website: www.epa.gov

**Federal Emergency Management Agency (FEMA)**
Main Number: 202.646.4600
Emergency Management Institute: 301.447.1000
National Emergency Training Center: 301.447.1048
National Fire Academy: 301.447.1000
Website: www.fema.gov

**Health and Human Services (HHS)**
Main Number: 202.690.7000
Office of Emergency Preparation: 202.566.1600
Website: http://ndms.dhhs.gov

# 10. The Future of Emergency Management

## INTRODUCTION

Philosopher George Santayana said "Those who cannot remember the past are condemned to repeat it." The discipline of emergency management is at a critical crossroad. Emergency managers are faced with new threats, new responsibilities, and new opportunities.

The potential for biochemical terrorist strikes, mass casualty events, and cyberspace attacks looms large. Providing protection to our first responders and to the general public from a myriad of unknown and unpredictable technological hazards is a daunting responsibility. Accepting this responsibility and wisely applying the lessons learned from emergency management practices and policies of the past represents both the challenge and the opportunity for emergency managers.

This chapter explores issues concerning the current political and organizational environment for emergency management. The chapter closes with the authors' opinions on what emergency management must do to survive and grow in this new environment.

## ORGANIZATIONAL CHANGES

After the events of September 11, the Bush administration and Congress focused on reorganization and increased appropriations to respond to the threat of terrorism. A new Department of Homeland Security has been created, which consolidates various federal agencies and programs with some responsibility for terrorism, including U.S. Border Patrol, Immigration and Naturalization Service, FEMA, Coast Guard, and a few other discrete programs. The reorganization does not include any of the intelligence, diplomatic, or law enforcement programs that are at the center of government efforts at preventing terrorism. The specific proposal for this department was discussed at length in Chapter 9. By including FEMA, the state and local emergency management structure of the United States will be integrated into the new department.

The second organizational move the Bush administration has taken is to propose establishing a new Northern Command as the backbone for homeland defense. The mission of the Northern Command is still being developed and the staffing of this Command (i.e., active military, active military reserves, and/or the National Guard) is still under discussion. What the mission becomes and the potential conflicts among the roles of these organizations have a major implication for emergency management.

Finally, the potential for significant new appropriations for terrorism and for state and local emergency management dominates the emergency management response to these reorganization initiatives. The U.S. emergency management system, at all levels, has been underfunded for decades.

In the early debate over the Nunn-Lugar antiterrorism legislation, emergency managers and other first responders, particularly the fire community, were lobbying for additional resources to prepare for possible terrorist attacks. It is unfortunate that during these discussions the fire and emergency management communities did not form a partnership to present a collective argument for their needs because it might have worked. Instead, the traditional rivalry between these two groups, both of which believe they are the most critical first responder, prevailed. The law enforcement community, on the other hand, presented a unified front. As a result, most of the Nunn-Lugar appropriated funds went to support the Department of Justice, FBI, and local law enforcement.

Later, the fire community was successful in establishing a new grant program to upgrade the deteriorating U.S. fire response infrastructure. The fire unions were responsible for getting these funds, and the terrorism threat was only one small part of their rationale. In the post-9/11 environment, it was obvious that funding for terrorism-related activities was going to be a high-priority competition.

The National Emergency Management Association (NEMA) and the International Association of Emergency Managers (IAEM) have endorsed the inclusion of FEMA in the new Department of Homeland Security. NEMA represents the State Directors of emergency management and IAEM, the locals. This endorsement comes although in most of the states, the governors have designated individuals other than the State Directors for Emergency Management as their agent or czar for Homeland Security. Other than the lure of money, it is hard to understand why the states would take this position. Because FEMA will lose stature and influence when it is no longer an independent agency and the Director of FEMA is no longer part of the President's Cabinet, so will the state emergency management organizations. There is no assurance that the proposed federal funding when sent to the states will be controlled by state emergency managers.

There is another aspect of these changes that the emergency managers need to consider. In many states, emergency management reports to the governor through the Adjutant General, who is the leader of the Army and Air National Guard. For years now, the National Guard has been looking for a new mission and new funding. As DOD budgets were reduced and state funding decreased, the Guard sought to expand its role in non–law enforcement events, particularly disasters where funding for their disaster support would be reimbursed by FEMA.

The involvement of the Guard in disaster response was not universally supported. While state emergency management offices grew in response to the natural disasters of the 1990s, National Guards retrenched. Preparedness for and response to terrorist events provides the Guard with the new mission they sought, and they are unlikely to relinquish it to the emergency managers. Furthermore, although it is unlikely that the nation's governors will ever give up their control of their State Guards, many people see the National Guard or at least the active military reserves as forming the backbone of the new Northern Command, giving them more money and authority through the DOD network. For example, California's Office of Homeland Security is working with the Adjutant General to implement a five-step state strategy on terrorism. One of the steps in this strategy is to reexamine state and federal legislation to see what needs to be changed to provide them with appropriate authority to operate in any emergency, technological or natural. This office believes it will have two missions: one in homeland security and one in homeland defense.

## WHAT DOES THIS MEAN FOR EMERGENCY MANAGEMENT?

It will probably take months, or more realistically years, to sort things out, but let's look at some of the potential changes. The implications for federal emergency management efforts are numerous. In the new Department of Homeland Security (DHS), FEMA becomes a division headed by an Undersecretary that reports up through a Deputy Secretary to the Department Secretary.

The direct authorities currently vested in the Director of FEMA will probably be given to the new DHS Secretary. Responsibilities for recommending disaster declarations to the President and for coordination of the federal response to natural and technological disasters or emergencies will probably be retained by the new DHS Secretary. This could dramatically impact the timeliness, effectiveness, and operational abilities of the current FEMA operations and staff. There may be changes in response or preparedness responsibilities to take better advantage of other parts of the new department, such as the Coast Guard. In any case, the stature and authorities of the leader of federal emergency management activities will be diminished.

Another likely impact will be the competition for resources among the various organizations within the new department. It is unlikely that the emergency management contingent will be effective in arguing for resources when up against organizations three and four times their size, such as the Immigration and Naturalization Services (INS). The increases in terrorism monies that are potentially flowing to emergency management can evaporate quickly in the absence of terrorist events or rescission of federal spending across the board. These impacts, if they are realized, will certainly extend to the states. They most assuredly will be felt at the local emergency management level, where we already see states using federal support designated for local efforts as offsets to state budget shortfalls.

Throughout the summer and fall of 2002, Congress debated various versions of the legislation that would create the Department of Homeland Security. It looked like the legislation might ultimately fail because of pressure from the Democrats in Congress to preserve the rights of Federal employees being transferred into the new Department. The mid-term elections of 2002 changed the control of Congress with a Republican majority in both houses. Without further deliberation on the legislation, the House passed the bill (H.R. 299-121) on November 13, 2002, and the Senate passed the bill (S.90-9) on November 19, 2002. On November 25, 2002, President Bush signed into law the Homeland Security Act of 2002 (P.L. 107-296) and announced that former Pennsylvania Governor Tom Ridge would be nominated to be Secretary of the new Department of Homeland Security established by the legislation. Ridge was quickly confirmed and promised that the new Department would be in place by March 1, 2003. The legislation allowed the Administration extraordinary freedom to reorganize personnel and programs from the existing 22 Agencies combined to create the new Department. For months a small group of individuals in the Office of Homeland Security was working on reorganization plans.

In January a small transition team headed by Secretary Ridge began to finalize the structure. The new structure moves significant numbers of FEMA personnel and most of the program funding for State and local emergency management infrastructure and first responders to a newly created Office of Domestic Preparedness. What remains of FEMA is a reduced personnel for disaster response and recovery, the U.S. Fire Administration, National Flood Insurance Program, other mitigation programs, non-terrorism training programs and Citizens Corps. Added into FEMA is the National Disaster Medical System, the National Strategic Stockpile and Nuclear Incident Response team. FEMA is renamed the Emergency Preparedness and Response Directorate.

On January 10, 2003, President Bush announced his intention to nominate Michael D. Brown to be the Under Secretary for Emergency Preparedness and Response. Mr. Brown had served as General Counsel and then Deputy Director to Joe Allbaugh, FEMA Director who had announced his resignation as Director of FEMA to be effective from March 1, 2003. Mr. Brown has stated that he would like to retain the name or the identification of "FEMA" even though it is not part of the official title of the new organization, since it has such universal public recognition and respect. This will be very difficult since a part of the name denotes an independent agency. Three events provide a level of insight into how the new Department of Homeland Security will operate and how it impacts emergency management and FEMA.

The organizations coming into the Department were taxed with providing initial operating funds for the new Department. FEMA was asked to provide approximately $35 million dollars, while the Coast Guard with a much larger budget was only required to contribute $3 million. FEMA came up with the funding by taking over half of the funding for the Flood Mitigation Assistance (FMA) program, funds from the pre-disaster mitigation grant program and the remaining from other accounts including the Disaster Relief Fund.

The second event was the tragic explosion of the Space Shuttle Columbia over Texas which was immediately declared a Federal Emergency using FEMA's authorities under the Stafford Act, as amended. This type of declaration is a prerogative of the President when Federal facilities or assets are impacted. The bombing of the Murrah Federal Building in Oklahoma City was initially declared a Federal emergency. The Columbia declaration took some by surprise as it was precedent-setting and had not been applied in the earlier Challenger disaster. Two speculations exist concerning this emergency. One is that the Bush Administration wanted to show the broad scope of the new Department and that it could perform its responsibilities. Another more practical reason may be that by declaring it an emergency, the Disaster Relief Fund could pay for the activities of the Agencies involved in the clean-up and investigation. Emergency declarations are capped at $5 million unless Congress is notified. As of February 24, FEMA has expended over $90 million responding to the Columbia shuttle explosion. The Disaster Relief Fund has historically been considered only applicable for victims, either people or communities. The leadership of the new Department may be taking a much broader approach to use of this Fund and it may provide further rationale for why FEMA was included in the new Department. While the Disaster Relief Fund receives an annual appropriation, most of its funding comes from supplemental appropriations that are outside of the normal Federal budget caps, thereby providing almost "free money." Congress has never failed to approve of supplemental funds in the aftermath of a single or multiple disasters.

Passage of the 2003 Omnibus Appropriations bill indicates that spending on federal level homeland security and terrorism programs is a high priority. However, heavy cuts were made in State and local homeland security programs such as grants for emergency management, first responders, and public safety. Overall resources to State and local were cut more than 35%.

Does each of these examples represent a special situation or are they representative of a significant change in philosophy? If the latter is true then this has broad implications for the future of emergency management at all levels of government. By rushing to adopt the new hazard of terrorism as the primary objective, emergency management may find itself losing the funding battle to other forces in law enforcement and defense.

## WHAT IS THE FUTURE OF EMERGENCY MANAGEMENT?

We are optimistic that emergency management can survive and thrive in the future if it embraces the lessons learned from the past and moves forward with a progressive agenda that will be valued by the American people.

Since September 11, the nation's psyche and its political leaders have been focused on terrorism and eliminating those who could perpetrate terrorist acts against the United States. At the same time, the nation has experienced relatively few major natural disasters. Large Western wildfires, many caused by human error and drought conditions, dominated the disaster landscape. The El Niño effect reduced the likelihood of floods and hurricanes. These conditions will change. Even when you consider the costs and human devastation of the September 11 events, the statistics indicate that natural disasters will continue to be costly. The probabilities support the likelihood of a major natural disaster, hurricane, or earthquake affecting our communities as inevitable. Continued development pressures will make flooding events in the United States more prevalent and severe. As emergency management systems focus their efforts on preparing for and responding to terrorist events, these efforts should not diminish their capabilities or capacity for dealing with natural hazards.

## Lesson One

*Maintain an all-hazards approach to emergency management.* Applying this approach takes advantage of the common capabilities necessary to treat any type of disaster or emergency but allows for incorporating the special needs of terrorism. To abandon the all-hazards approach would be repeating the mistake the emergency management community made in the 1980s. During the era of the cold war, FEMA concentrated more than 75 percent of its financial and human resources on preparing for the next nuclear war. It mandated that states and localities receiving FEMA funding follow suit.

Federal, state, and local capacity to respond to natural disasters was severely diminished. As Hurricanes Hugo, Iniki, and Andrew vividly demonstrated, state and local capacities were quickly overcome. The federal response under FEMA was disorganized and late. In the case of Andrew, the Director of FEMA was replaced as the in-charge official and the military provided most of the initial support. This real example of the folly of focusing on any one threat, at the cost of more frequent and widespread threats, provides strong evidence of the wisdom of the all-hazards approach to emergency management.

## Lesson Two

*The federal response infrastructure, based on the Federal Response Plan, works.* Since September 11, many political leaders have called for building a terrorism response structure, forgetting that an effective federal structure already exists. There is no need to build a new infrastructure. This approach was tested in hundreds of natural events and the Oklahoma City bombing—a terrorist event—and it worked. This proven structure is flexible; it needs modification and the addition of new partners to accommodate the unique aspects of terrorism, but the emergency management community should fight any attempts to build a separate structure.

At the state and local levels, state plans and the emergency management compacts that exist between states support this operational approach. Specific lessons learned from September 11, particularly in communications and joint operations, can be readily incorporated into these existing structures.

## Lesson Three

*Continue to practice the concepts that facilitated the U.S. emergency management system becoming the best system in the world.* These concepts are (1) focus on your customers, both internal and external; (2) build partnerships among disciplines and across sectors, including the private sector and the media; (3) support development and application of new technologies to give emergency managers the tools they need to be successful; (4) emphasize communications to partners, the public, and the media; and (5) make prevention the cornerstone of emergency management.

These simple, common sense concepts were the key to the respect and success FEMA achieved under Director James Lee Witt and President Clinton. We believe they provide the framework for emergency management to continue to grow and expand its influence and importance to the institutions and people it serves. Emergency management can ensure its place in the future if it focuses on policies, programs, and activities that improve the safety and social and economic security of individuals, institutions, and communities. To do this, emergency management must focus more effort in promoting and implementing prevention and mitigation.

## PREVENTION AND MITIGATION

Prevention is the positive function that emergency managers can practice everyday, in every community, and not be dependent on an event to prove their value. Prevention is practiced by all sectors of a community. To be effective, it requires developing partnerships within a community and often brings together disparate parties to solve common problems. Mitigation brings the private sector into the emergency management system because economic sustainability of their businesses depends on risk reduction, so prevention promotes their support and leadership. Mitigation provides the entry point to involve the private sector in other phases of emergency management and to understand their unique needs in response and recovery.

In the late 1990s, business continuity and mitigation planning was the largest growth area for emergency management. Economic considerations or interest often drive public decisions. Mitigation allows emergency managers to have access and influence to the decision-making process. Mitigation works best at the local level and provides that grassroots constituency that can exert political pressure for continued emergency management support. The Project Impact initiative articulated this concept and made it a reality in more than 250 communities. The Bush administration recognized this by including the words "building disaster-resistant communities" in the objectives for the new Department of Homeland Security.

There are many competitors for the role of homeland security czar. There are reasons and politics that may affect who or what agency gains prominence. While the struggle goes on, the emergency management community can demonstrate their value by focusing on mitigation. Their investment in this approach will pay off. Should another terrorism event happen, the same questions raised after September 11 will surface again because no preparedness or response will be adequate; however, when the next natural hazard occurs, emergency management leadership in prevention and mitigation will be recognized and rewarded with public support. If there is any doubt, the events and comments made by public officials and citizens in the aftermath of the 2001 Seattle earthquake prove the point.

## CONCLUSION

Whether the emergency management establishment will embrace this path in the future is debatable. Historic trends indicate otherwise; however, throughout the

1990s a new breed of emergency management professionals began to emerge. These individuals were anxious to bring a fresh face to the profession and embraced new strategies for promoting sound emergency management practices, particularly mitigation. The future of emergency management may rest on their ability to balance the new demands of the terrorism threat with the real need to make a difference in the quality of people's lives and their community's sustainability through mitigation.

# References

"After-Action Report on the Response to the September 11 Terrorist Attack of the Pentagon." Washington, DC: Titan Systems Inc., July 2002.

American Red Cross Disaster Operations Center Visitor's Guide.

Bakhet, Omar. *Linking Relief to Development*. UNDP Rwanda, June 1998.

Bagli, Charles V. "Seeking Safety, Downtown Firms Are Scattering." *The New York Times,* January 29, 2002, p. A-1.

Burke, Robert. *Counter Terrorism for Emergency Responders*. CRC/Lewis Publishers, 2000.

Center for Disaster Management and Humanitarian Assistance. *NGOs and Disaster Response Who Are These Guys and What Do They Do Anyways?,* (n.d.), www.cdmha.org/ppt/%20presentation.ppt.

Coyle, David. *The United Nations and How it Works*. New York: Columbia University Press, 1969.

Department of Defense, CCRP. *The Complex Process of Responding to Crisis*, (n.d.), www.dodccrp.org/ngoCh2.html.

Emergency Management Institute (EMI). 2001–2002 Catalog of Activities, Emmitsburg, MD. www.fema.gov.

Erickson, Paul A. *Emergency Response Planning for Corporate and Municipal Managers*. Federal Emergency Management Agency. Report on Costs and Benefits of Natural Hazard Mitigation. Washington, DC: FEMA, March 1997.

Federal Emergency Management Agency. *Partnerships in Preparedness, A Compendium of Exemplary Practices in Emergency Management, Volume IV*. Washington, DC: FEMA, January 2000.

Federal Emergency Management Agency. *Partnerships in Preparedness, A Compendium of Exemplary Practices in Emergency Management, Volume II*. Washington, DC: FEMA, May 1997.

Federal Emergency Management Agency. *International Technical Assistance Activities of the United States Federal Emergency Management Agency*. Washington, DC: FEMA, 2001.

Federal Emergency Management Agency. *Federal Response Plan*. Washington, DC: FEMA, April 1999.

Federal Emergency Management Agency, Office of the Inspector General. *FEMA's Disaster Management Program: A Performance Audit After Hurricane Andrew*. H-01-93. Washington, DC: FEMA, January 1993.

*FEMA Emergency Information Field Guide* (condensed). October 1998.

FEMA Website, www.fema.gov.

Flynn, Kevin. "After the Attack: The Firefighters; Department's Cruel Toll: 350 Comrades." *New York Times*, September 13, 2001, Section: National Desk.

Gandel, Stephen. "Consultants push Wall Street to leave; Downtown's losses are huge, but some companies shrug off fears, concentrate workers in midtown." *Crain's New York Business* March 4, 2002, p. 1.

Gilbert, Alorie. "Out of the Ashes." *Information Week*, March 2002. [Online], 3 pages, www.informationweek.com/story/IWK20020104S0008.

Gilbert, Alorie. (2002, March). *Out of the Ashes*. InformationWeek [Online], 3 pages. Available: http://www.informationweek.com/story/IWK20020104S0008 [2002, March 14].

Gilbert, Roy, and Alcira Kreimer. *Learning from the World Bank's Experience of Natural Disaster Related Assistance*. Washington, DC: The World Bank, 1999.

GlobalCorps. *OFDA's Evolving Role*, (n.d.), www.globalcorps.com/ofda/ofdarole.html.

Hawley, Chris. "Globalization and Sept. 11 are pushing Wall Street off Wall Street, analysts say." The Associated Press State & Local Wire, February 1 2002, State and Regional

Section. *Lexis-Nexis Universe: World Trade Center, Firm and Tenant*. Online. March 20, 2002.

Hedges, Chris. "Monday Counting Losses, Department Rethinks Fighting Every Fire." *New York Post*, October 1, 2001.

Houston, Allen. "Crisis Communications: Confused Messages Spur Switch to Sole Spokesman." *PR Week*, November 5, 2001, p. 11.

Houston, Allen. "Crisis Communications: The Readiness is All: How the Red Cross Responded." *PR Week*, October 22, 2001, p. 13.

Intergovernmental Panel on Climate Change. *Special Issues in Developing Countries*, 2001, www.ipcc.ch/pub/tar/wg2/340.htm.

International Monetary Fund. *IMF Emergency Assistance Related to Natural Disasters and Post-Conflict Situations: A Factsheet*, 2001, www.imf.org/external/np/eexr/facts/conflict.htm.

Irwin, Robert L. (1980) *Disaster Response: Principles of Preparation and Coordination.* Chapter 7: The Incident Command System (ICS).

Madison County, North Carolina. "Multi-Hazard Plan for Madison County."

Mancino, Kimberly. *Development Relief: NGO Efforts to Promote Sustainable Peace and Development in Complex Humanitarian Emergencies*. Washington, DC: Interaction, 2001.

Maynard, Kimberly. *Healing Communities in Conflict: International Assistance in Complex Emergencies*, (n.d.), www.ciaonet.org/book/maynard/maynard07.html.

McKinsey & Company. "Improving NYPD Emergency Preparedness and Response," August 19, 2002.

Mileti, Dennis S. *Disasters by Design: A Reassessment of Natural Hazards in the United States*. Washington, DC: John Henry Press, 1999.

Mitchell, James K., et al. *Crucibles of Hazard: Mega-Cities and Disasters in Transition.* New York: United Nations University Press, 1999.

National Association of Counties. "Counties and Homeland Security: Policy Agenda to Secure the People of America's Counties." August 2002 [online], www.naco.org/programs/homesecurity?policyplan.cfm.

National Emergency Management Association. "NEMA Reports on State Homeland Security Structures," June 2002 [online], www.nemaweb.org/ShowExtendedNewscfm?ID = 171.

National Emergency Management Association. "White Paper on Domestic Preparedness." October 1, 2001.

Natsios, Andrew S. *U.S. Foreign Policy and the Four Horsemen of the Apocalypse*. Westport, CT: Praeger Publishers, 1997.

Office of Homeland Security. "State and local Actions for Homeland Security, July 2002 [online], www.whitehouse.gov/homeland/stateandlocal.

Office for the Coordination of Humanitarian Affairs. *Information Summary on Military and Civil Defence Assets (MCDA) and the Military and Civil Defence Unit (MCDU)*, (n.d.), www.reliefweb.int/ocha_ol/programs/response/mcdunet/0mcduinf.html.

Office for the Coordination of Humanitarian Affairs. *Coordination of Humanitarian Response*, (n.d.), www.reliefweb.int/ocha_ol/programs/response/service.html.

Otero, Juan. "Congress, Administration Examines Emergency Communications Systems," *Nation's Cities Weekly*, October 22, 2001, p. 5.

Pan American Health Organization. *Natural Disasters: Protecting the Public's Health*, (n.d.) PAHO Scientific Publication No. 575.

Powell, Michael, and Christine Haughney. "A Towering Task Lags in New York, City Debates Competing Visions for Rebuilding Devastated Downtown." The Washington Post.com [Online], Page A03. Available: www.washingtonpost.com/ac2/.../A22040-2002Feb16?language=printe [2002, February 25].

Ranganath, Priya. *Mitigation and the Consequences of International Aid in Postdisaster Reconstruction*. Centre d'Etude et de Cooperation Internationale, 2000.

Rendleman, John. "Back Online, Despite its losses, Verizon went right back to work restoring communications services." October 29, 2001. InformationWeek.com [Online], 2 pages. Available: www.informationwee.com/story/IWK2001105S0017.

Salomons, Dirk. *Building Regional and National Capacities for Leadership in Humanitarian Assistance*. New York: The Praxis Group, 1998.

Site One. *History of the Red Cross*, (n.d.), www.siteone.com/redcross/anniv.htm.

Smith, Lloyd R. "Lessons Learned from Oklahoma City: Your Employees . . . Their Needs, Their Role in Response and Recovery." www.recovery.com.

Stock, Gary. (2002, October 14). *List of Business Offices, Tenants, and Companies in the World Trade Center (WTC)*. UnBlinking.com [Online], 28 pages. Available: http://www.tbtf.com/unblinking/are/2001/-09a.htm [2002, March 4].

The Oklahoma Department of Civil Emergency Management. "After Action Report, Alfred P. Murrah Building Bombing. Lessons Learned." 1997.

"The Look Back at Mitch's Rampage," *USA Today*, June 8, 1999, www.usatoday.com/weather/news/1998/wmrmpage.htm.

The United States Conference of Mayors. "A National Action Plan for Safety and Security in America's Cities." December 2001. www.usmayors.org/uscm/home.asp.

The World Bank. "Assistance to Post-Conflict Countries and the HIPC Framework." April 2000, www.imf.org/external/np/hipc/2001/pc/042001.htm.

Town of Boone, North Carolina. "All Hazards Planning and Operations Manual," July 1999.

"Tsunami Education a Priority in Hawaii and West Coast States." *Bulletin of the American Meteorological Society*, June 2001, p. 1207.

U.S. Agency for International Development. *Field Operations Guide for Disaster Assessment and Response, Version 3.0*. Washington, DC: USAID, 1998.

U.S. Agency for International Development. *OFDA Annual Report 2000*, (n.d.), Washington, DC: USAID, 2000.

U.S. Agency for International Development. "Rebuilding Postwar Rwanda: The Role of the International Community." 1996, www.usaid.gov/pubs/usaid_eval/ascii/pnaby212.txt.

U.S. Department of State. "Bureau of Population, Refugees, and Migration," (n.d.), www.state.gov/g/prm.

UN Rwanda. "The United Nations Development Programme in Rwanda," (n.d.), www.un.rw/UNRwa/UNDP.shtml.

UNDP. "About ERD," (n.d.), www.undp.org/erd/about.htm.

UNDP. "Building Bridges Between Relief and Development: A Compendium of the UNDP in Crisis Countries," (n.d.), www.undp.org/erd/archives/bridges.htm.

UNDP. *Disaster Profiles of the Least Developed Countries*. United Nations, May 2001.

UNDP. "Further Elaboration on Follow-Up to Economic and Social Council Resolution 1995/96: Strengthening of the Coordination of Emergency Humanitarian Assistance," Second Regular Session, March 10–14 1997, www.undp.org/erd/archives/executiv.htm.

UNDP. "Reintegration & Rehabilitation," (n.d.), www.undp.org/erd/randr.htm.

UNDP. "Strengthening National Disaster Management," (n.d.), www.undp.org/erd/archives/brochures/drrp/countries/Botswana.htm.

UNDP. "The Disaster Reduction and Recovery Program: An Introduction," (n.d.), www.undp.org/erd/archives/brochures/drrp/main.htm.

UNDP. "The United Nations Development Programme Mission Statement," (n.d.), www.undp.org/info/discover/mission.html.

UNDP. "Working for Solutions to Crises: The Development Response," (n.d.), www.undp.org/erd/archives/bridges2.htm.

United Nations. General Assembly Economic and Social Council, United Nations, May 8, 2001.

U.S. Army. "About NGOs," (n.d.), http://call.army.mil/fmso/ngos/introduction.html.

"VIPs, Disaster Service Calls May Get Priority." *The Daily Yomiuri* (Tokyo), The Yomiuri Shimbun, November 26, 2001, p. 1.

Walsh, Edward. "National Response to Terror; FEMA leads Effort; Borders Tightened." *The Washington Post*, September 12, 2001, p. A-1.

Waugh, William Jr. *Living with Hazards—Dealing with Disasters: An Introduction to Emergency Management*. New York: M.E. Sharpe, 2000.

Wax, Alan J., and Diop, Julie Claire. "Return to Downtown; Office leases are being signed again, but revival will take a while." *Newsday*, March 11, 2002, Business and Technology Section: p. D13. *Lexis-Nexis Universe: World Trade Center, Firm and Tenant*. Online. 15 Mar. 2002.

"Wireless System Improves Communications," *American City & County*, August 2001.

Zevin, Rona. "Tapping Web Power in Emergencies," *American City & County*, August 2001.

# Appendix A

## Acronyms

| | |
|---|---|
| AAR | After-Action Report |
| AEC | Agency Emergency Coordinator |
| AFRO | African Regional Office (WHO) |
| AOA | Administration on Aging |
| AOR | Areas of Responsibility (DOD) |
| ARC | American Red Cross |
| ARES | Amateur Radio Emergency Services |
| BHR | Bureau for Humanitarian Response (USAID) |
| CARE | Cooperative for Assistance and Relief Everywhere |
| CAT | Crisis Action Team |
| CDBG | Community Development Block Grant |
| CCP | Crisis Counseling Assistance and Training Program |
| CCP | Casualty Collection Point |
| CDC | Centers for Disease Control and Prevention, U.S. Public Health Service |
| CDRG | Catastrophic Disaster Response Group |
| CENTCOM | Central Command (DOD) |
| CEPPO | Chemical Emergency Preparedness and Prevention Office |
| CERCLA | Comprehensive Environmental Response, Compensation and Liability Act |
| CFDA | Catalog of Federal Domestic Assistance |
| CHE | Complex Humanitarian Emergency |
| CJTF | Commander for the Joint Task Force (DOD) |
| CMHS | Center for Mental Health Services |
| CMOC | Civil/Military Operations Center (DOD) |
| CMT | Crisis Management Team |
| CNN | Cable News Network |
| CRC | Convention on the Rights of the Child |
| CRC | Crisis Response Cell |
| CRM | Crisis Resource Manager |
| CRS | Catholic Relief Services |
| DAE | Disaster Assistance Employee |
| DART | Disaster Assistance Response Team (USAID) |
| DEA | Drug Enforcement Agency |
| DFO | Disaster Field Office |
| DHHS | Department of Health and Human Services |
| DMAT | Disaster Medical Assistance Team |

| | |
|---|---|
| DMORT | Disaster Mortuary Response Team, National Disaster Medical System |
| DMTP | Disaster Management Training Programme |
| DOD | United States Department of Defense |
| DOJ | Department of Justice |
| DOL | Department of Labor |
| DOT | Department of Transportation |
| DRC | Disaster Recovery Center |
| DRD | Disaster Response Division |
| DRRP | Disaster Reduction and Recovery Programme |
| DUA | Disaster Unemployment Assistance |
| EAS | Emergency Alert System |
| EC | Emergency Coordinator |
| ECHO | European Community Humanitarian Organization |
| ECS | Emergency Communications Staff |
| EDA | Economic Development Administration |
| EGOM | Empowered Group of Ministers (India) |
| EICC | Emergency Information and Coordination Center |
| EMRO | Eastern-Mediterranean Regional Office (WHO) |
| EMS | Emergency Medical Services |
| EOC | Emergency Operations Center |
| ERC | Emergency Response Coordinator (UN) |
| ERCG | Emergency Response Coordination Group, Public Health Service/Centers for Disease Control and Agency for Toxic Substances and Disease Registry |
| ERD | Emergency Response Division (UNDP) |
| ERL | Emergency Recovery Loan (WBG) |
| ERT | Emergency Response Team |
| ERT-A | Advance Element of the Emergency Response Team |
| ERU | Emergency Response Unit (IFRC) |
| ESF | Emergency Support Function |
| EST | Emergency Support Team |
| EUCOM | European Command (DOD) |
| EURO | Regional Office for Europe (WHO) |
| FAA | Federal Aviation Administration |
| FACT | Field Assessment and Coordination Team (IFRC) |
| FAO | Food and Agriculture Organization |
| FBI | Federal Bureau of Investigation |
| FCO | Federal Coordinating Officer |
| FECC | Federal Emergency Communications Coordinator |
| FEMA | Federal Emergency Management Agency |
| FERC | FEMA Emergency Response Capability |
| FESC | Federal Emergency Support Coordinator |
| FFP | Office of Food For Peace (BHR) |
| FHA | Foreign Humanitarian Assistance (DOD) |
| FHWA | Federal Highway Administration |

| | |
|---|---|
| FOC | FEMA Operations Center |
| FRERP | Federal Radiological Emergency Response Plan |
| FRN | FEMA Radio Network |
| FRP | Federal Response Plan |
| FSA | Farm Service Agency |
| HAO | Humanitarian Assistance Operations (DOD) |
| HAST | Humanitarian Assistance Survey Team (DOD) |
| HET-ESF | Headquarters Emergency Transportation Emergency Support Function |
| HHS | Department of Health and Human Services |
| HUD | Department of Housing and Urban Development |
| IAEM | International Association of Emergency Managers |
| IASC | Inter-Agency Standing Committee |
| IBRD | International Bank for Reconstruction and Development (WBG) |
| ICPAE | Interagency Committee on Public Affairs in Emergencies |
| ICRC | International Committee of the Red Cross |
| ICS | Incident Command System |
| ICVA | International Council for Voluntary Agencies |
| IDA | International Development Association (WBG) |
| IDNDR | International Decade for Natural Disaster Reduction (UN) |
| IDP | Internally Displaced Persons |
| IFC | International Finance Corporation (WBG) |
| IFG | Individual and Family Grant |
| IFRC | International Federation of Red Cross/Red Crescent Societies |
| IMD | Indian Meteorological Department |
| IMF | International Monetary Fund |
| INS | Immigration and Naturalization Service |
| IO | International Organization |
| ICSID | International Centre for Settlement of Investment Disputes (WBG) |
| ISDR | International Strategy for Disaster Reduction (UN) |
| JCS | Joint Chiefs of Staff (DOD) |
| JIC | Joint Information Center |
| JTF | Joint Task Force (DOD) |
| MIGA | Multilateral Investment Guarantee Agency (WBG) |
| MOA | Memorandum of Agreement |
| MOU | Memorandum of Understanding |
| MSF | Medecin sans Frontiers |
| NACo | National Association of Counties |
| NASA | National Aeronautics and Space Agency |
| NCA | National Command Authority (DOD) |
| NDMOC | National Disaster Medical Operations Center |
| NDMS | National Disaster Medical System |
| NDMSOSC | National Disaster Medical System Operations Support Center |
| NEIC | National Earthquake Information Center |
| NEMA | National Emergency Management Association |

| | |
|---|---|
| NEPEC | National Earthquake Prediction Evaluation Council |
| NGO | Nongovernmental Organization |
| NOAA | National Oceanic and Atmospheric Administration |
| NPSC | National Processing Service Center |
| NRC | Nuclear Regulatory Commission |
| NRT | National Response Team |
| NSEP | National Security Emergency Preparedness |
| NVOAD | National Voluntary Organizations Active in Disaster |
| OCHA | Office for the Coordination of Humanitarian Affairs |
| OEP | Office of Emergency Preparedness, U.S. Public Health Service |
| OET | Office of Emergency Transportation |
| OFDA | Office of U.S. Foreign Disaster Assistance |
| OPA | Office of Public Affairs |
| OS | Operation Support (OFDA) |
| OSC | On-Scene Coordinator |
| OTI | Office of Transition Initiatives (BHR) |
| PACOM | Pacific Command (DOD) |
| PAHO | Pan-American Health Organization (WHO) |
| PAO | Public Affairs Officer |
| PK/HA | Office of Peacekeeping and Humanitarian Affairs (DOD) |
| PM | Office of Political/Military Affairs (DOD) |
| PMPP | Prevention, Mitigation, Preparedness, and Planning (OFDA) |
| PNP | Private Nonprofit |
| PRM | Bureau of Population, Refugees and Migration (USAID) |
| PS | Program Support (OFDA) |
| PSA | Public Service Announcement |
| PSYOPS | Psychological Operations (DOD) |
| PVO | Private Voluntary Organization |
| QIP | Quick Impact Project (UNHCR) |
| RACES | Radio Amateur Civil Emergency Services |
| REACT | Radio Emergency Associated Communication Team |
| REC | Regional Emergency Coordinator |
| RECC | Regional Emergency Communications Coordinator |
| RECP | Regional Emergency Communications Plan |
| RET | Regional Emergency Transportation |
| RETCO | Regional Emergency Transportation Coordinator |
| RMT | Response Management Team (OFDA) |
| ROC | Regional Operations Center |
| ROE | Rules of Engagement (DOD) |
| ROST | Regional Operations Support Team |
| RRT | Regional Response Team |
| SAMHSA | Substance Abuse and Medical Health Services Administration |
| SAR | Search and Rescue |
| SBA | U.S. Small Business Administration |
| SCO | State Coordinating Officer |
| SEARO | South-East Asia Regional Office (WHO) |

| | |
|---|---|
| SFHA | Special Flood Hazard Areas |
| SITREP | Situation Report |
| SOCOM | Special Operations Command (DOD) |
| SOUTHCOM | Southern Command (DOD) |
| TAG | Technical Assistance Group (OFDA) |
| TRANSCOM | Transportation Command (DOD) |
| UN | United Nations |
| UNDAC | UN Disaster Assessment and Coordination |
| UNDP | United Nations Development Programme |
| UNFPA | United Nations Populations Fund |
| UNHCR | United Nations High Commissioner for Refugees |
| UNHRD | UN Humanitarian Response Depot |
| UNICEF | United Nations Children's Fund |
| US&R/USAR | Urban Search and Rescue |
| USACE | United States Army Corps of Engineers |
| USACOM | United States Atlantic Command (DOD) |
| USAID | United States Agency for International Development |
| USDA | United States Department of Agriculture |
| USGS | United States Geological Survey |
| WB | World Bank |
| WBG | World Bank Group |
| WFP | World Food Programme |
| WHO | World Health Organization |
| WTC | World Trade Center |
| ZECP | Zone Emergency Communications Planner |

# Appendix B

## Emergency Management Websites

| Emergency Type | Contact Agency | Website |
|---|---|---|
| All hazards | The Federal Emergency Management Agency (FEMA) | www.fema.gov |
| Hazardous materials, chemical accidents, oil spills | Environmental Protection Agency (EPA) | www.epa.gov/ceppo |
| | National Response Center | www.nrc.uscg.mil |
| Land-based natural hazards, earthquakes, floods, volcanoes | U.S. Geological Survey (USGS) | www.usgs.gov |
| Earthquakes | National Earthquake Information Center (NEIC) | www.neic.cr.usgs.gov |
| | Earthquake Engineering Research Institute | www.eeri.org |
| Floods | Association of State Floodplain Managers | www.floods.org |
| Hurricanes and meteorological hazards | National Oceanographic and Atmospheric Administration (NOAA) | www.nhc.noaa.gov/ www.noaa.gov |
| Wildland fires | National Interagency Fire Center | www.nifc.gov |
| National security hazard information | American Red Cross | www.redcross.org |
| | Department of Health and Human Services (DHHS): Anthrax and Biological Incidents | www.hhs.gov/hottopics/healing/biological.html |
| | Department of Transportation (DOT) | http://ntl.bts.gov/faz/sept11.html |
| | Environmental Protection Agency (EPA) | www.epa.gov/safewater |
| | Federal Bureau of Investigation (FBI) | www.fbi.gov/pressrel/attack/attacks.htm |
| | Federal Consumer Information Center | www.pueblo.gsa.gov/crisis.htm |
| | Office of Homeland Security (OHS) | www.whitehouse.gov/homeland |
| | Small Business Administration Economic Injury Disaster Loans | www.sba.gov/news/current01/economicinjuryfactsheet.html |
| | U.S. Postal Service Updates | www.usps.com/news/2001/press/serviceupdates.htm |
| | White House Federal Recovery Action | www.whitehouse.gov/response/fedresponse.html |
| Disaster services | American National Red Cross | www.redcross.org |
| | Disaster Relief | www.disasterrelief.org |
| | Pacific Disaster Center (Information Technology for Disaster Response) | www.pdc.gov |

*continued*

| Emergency Type | Contact Agency | Website |
|---|---|---|
| | National Voluntary Organizations Active in Disaster (NVOAD) | www.nvoad.org |
| State and local organizations | National Emergency Management Association (NEMA)—State Emergency Managers Association | www.nemaweb.org |
| | International Association of Emergency Managers (IAEM)—Local Emergency Managers Association | www.iaem.com |
| | Extension Disaster Education Network | www.agctr.lsu.edu/eden |
| | University of Colorado National Hazards Center | www.colorado.edu/hazards |
| | University of Delaware Disaster Research Center | www.udel.edu/DRC |
| Federal government websites | Health and Human Services Administration on Aging | www.aoa.dhhs.gov/default.htm |
| | Health and Human Services (DHHS)—Center for Mental Health Services | www.mentalhealth.org/cmhs/EmergencyServices/default.asp |
| | Department of Commerce—Economic Development Administration (EDA) | www.doc.gov/eda |
| | Department of Labor (DOL)—Disaster Unemeployment Assistance | http://workforcesecurity.doleta.gov/unemploy/disaster.asp |
| | Department of Transportation (DOT)—Federal Highway Administration (FHA) | http://workforcesecurity.doleta.gov/unemploy/disaster.asp |
| | Department of Housing and Urban Development (HUD) | www.hud.gov/disassit.cfm |
| | Small Business Administration (SBA) | www.sba.gov/disaster |
| | U.S. Army Corps of Engineers | www.usace.army.mil/index.html |
| | U.S. Department of Agriculture (USDA)—Farm Service Agency (FSA) | www.fsa.usda.gov/pas/default.asp |
| International websites | International Federation of Red Cross/Red Crescent Societies | www.ifrc.org |
| | International Committee of the Red Cross | www.icrc.org |
| | United Nations Children's Fund | www.unicef.org |
| | International Monetary Fund (IMF) | www.imf.org |
| | The World Bank | www.worldbank.org |
| | United Nations Development Programme (UNDP) | www.undp.org |
| | International Strategy for Disaster Reduction (UN) | www.unisdr.org |
| | World Food Programme (WFP) | www.wfp.org |
| | World Health Organization (WHO) | www.who.org |
| | United Nations High Commissioner for Refugees | www.unhcr.ch |
| | U.S. Agency for International Development (USAID) | www.usaid.gov |
| | U.S. Department of Defense (DOD) | www.dod.mil |
| | Interaction | www.interaction.org |

# Appendix C

## Emergency Management Agency Addresses

**FEMA HEADQUARTERS**

500 C Street SW
Washington, DC 20472
(Note: Please use this address to reach the following offices by inserting the Office Name as the second line of the address:)

- Office of the Director
- Office of National Security Affairs
- Office of Intergovernmental Affairs
- Office of Policy and Regional Operations
- Office of the General Counsel
- Office of Public Affairs
- Office of Human Resources
- Office of Equal Rights
- Office of Financial Management
- Office of the Inspector General
- Mitigation Directorate
- Preparedness, Training and Exercises Directorate
- Response and Recovery Directorate
- Federal Insurance Administration
- Operations Support Directorate
- Information Technology Directorate

**National Emergency Training Center**
16825 South Seton Avenue
Emmitsburg, MD 21727

**U.S. Fire Administration**
16825 South Seton Avenue
Emmitsburg, MD 21727

**Emergency Management Institute**
16825 South Seton Avenue
Emmitsburg, MD 21727

**Mount Weather Emergency Assistance Center**
19844 Blue Ridge Mountain Road
State Route 601
Bluemont, VA 20135

## REGIONAL OFFICES

**FEMA Region I**
442 J.W. McCormack POCH
Boston, MA 02109-4595
(Note: This office serves the states of Maine, New Hampshire, Vermont, Rhode Island, Connecticut, and the Commonwealth of Massachusetts.)

**FEMA Region II**
26 Federal Plaza, Room 1337
New York, NY 10278-0002
(Note: This office serves the states of New York, New Jersey, the Commonwealth of Puerto Rico, and the Territory of the U.S. Virgin Islands.)

**FEMA Region III**
615 Chestnut Street
One Independence Mall, Sixth Floor
Philadelphia, PA 19106-4404
(Note: This office serves the states of Delaware, Maryland, Pennsylvania, Virginia, West Virginia, and the District of Columbia.)

**FEMA Region IV**
3003 Chamblee Tucker Road
Atlanta, GA 30341
(Note: This office serves the states of Alabama, Florida, Georgia, Kentucky, Mississippi, North Carolina, South Carolina, and Tennessee.)

**FEMA Region V**
536 South Clark St., 6th Floor
Chicago, IL 60605
(Note: This office serves the states of Illinois, Indiana, Michigan, Minnesota, Ohio, and Wisconsin.)

**FEMA Region VI**
FRC 800 North Loop 288
Denton, TX 76209
(Note: This office serves the states of Arkansas, Louisiana, New Mexico, Oklahoma, and Texas.)

**FEMA Region VII**
2323 Grand Boulevard, Suite 900
Kansas City, MO 64108-2670
(Note: This office serves the states of Iowa, Kansas, Missouri, and Nebraska.)

**FEMA Region VIII**
Denver Federal Center
Building 710, Box 25267
Denver, CO 80255-0267
(Note: This office serves the states of Colorado, Montana, North Dakota, South Dakota, Utah, and Wyoming.)

**FEMA Region IX**
1111 Broadway, Suite 1200
Oakland, CA 94607
(Note: This office serves the states of Arizona, California, Hawaii, and Nevada; and the Territory of American Samoa, the Territory of Guam, the Commonwealth of the Northern Mariana Islands, the Republic of the Marshall Islands, the Federated States of Micronesia, and the Republic of Palau.)

**FEMA Region X**
Federal Regional Center
130 228th Street, SW
Bothell, WA 98021-9796
(Note: This office serves the states of Alaska, Idaho, Oregon, and Washington.)

## STATE OFFICES AND AGENCIES OF EMERGENCY MANAGEMENT

**Alabama Emergency Management Agency**
5898 County Road 41
P.O. Drawer 2160
Clanton, AL 35046-2160
(205) 280-2200
(205) 280-2495 fax
www.aema.state.al.us

**Alaska Division of Emergency Services**
P.O. Box 5750
Fort Richardson, AK 99505-5750
(907) 428-7000
(907) 428-7009 fax
www.ak-prepared.com

**American Samoa Territorial Emergency Management Coordination (TEMCO)**
American Samoa Government
P.O. Box 1086
Pago Pago, American Samoa 96799
(011)(684) 699-6415
(011)(684) 699-6414 fax

**Arizona Division of Emergency Services**
5636 East McDowell Road
Phoenix, AZ 85008
(602) 231-6245
(602) 231-6356 fax
www.state.az.us/es

**Arkansas Department of Emergency Management**
P.O. Box 758
Conway, AR 72033

(501) 730-9750
(501) 730-9754 fax
www.adem.state.ar.us

**California Governor's Office of Emergency Services**
P.O. Box 419047
Rancho Cordova, CA 95741-9047
(916) 845-8510
(916) 845-8511 fax
www.oes.ca.gov

**Colorado Office of Emergency Management**
Division of Local Government
Department of Local Affairs
15075 South Golden Road
Golden, CO 80401-3979
(303) 273-1622
(303) 273-1795 fax
www.dola.state.co.us/oem/oemindex.htm

**Connecticut Office of Emergency Management**
Military Department
360 Broad Street
Hartford, CT 06105
(860) 566-3180
(860) 247-0664 fax
www.mil.state.ct.us/OEM.htm

**Delaware Emergency Management Agency**
165 Brick Store Landing Road
Smyrna, DE 19977
(302) 659-3362
(302) 659-6855 fax
www.state.de.us/dema/index.htm

**District of Columbia Emergency Management Agency**
2000 14th Street NW, 8th Floor
Washington, DC 20009
(202) 727-6161
(202) 673-2290 fax
www.dcema.dc.gov

**Florida Division of Emergency Management**
2555 Shumard Oak Blvd.
Tallahassee, FL 32399-2100
(850) 413-9969
(850) 488-1016 fax
www.floridadisaster.org

**Georgia Emergency Management Agency**
P.O. Box 18055
Atlanta, GA 30316-0055
(404) 635-7000
(404) 635-7205 fax
www.State.Ga.US/GEMA

**Office of Civil Defense, Government of Guam**
P.O. Box 2877
Hagatna, Guam 96932
(011)(671) 475-9600
(011)(671) 477-3727 fax
http://ns.gov.gu

**Hawaii State Civil Defense**
3949 Diamond Head Road
Honolulu, HI 96816-4495
(808) 734-4246
(808) 733-4287 fax
http://scd.state.hi.us

**Idaho Bureau of Disaster Services**
4040 Guard Street, Bldg. 600
Boise, ID 83705-5004
(208) 334-3460
(208) 334-2322 fax
www.state.id.us/bds/bds.html

**Illinois Emergency Management Agency**
110 East Adams Street
Springfield, IL 62701
(217) 782-2700
(217) 524-7967 fax
www.state.il.us/iema

**Indiana State Emergency Management Agency**
302 West Washington Street
Room E-208 A
Indianapolis, IN 46204-2767
(317) 232-3986
(317) 232-3895 fax
www.ai.org/sema/index.html

**Iowa Division of Emergency Management**
Department of Public Defense
Hoover Office Building
Des Moines, IA 50319
(641) 281-3231
(641) 281-7539 fax
www.state.ia.us/government/dpd/emd/index.htm

**Kansas Division of Emergency Management**
2800 S.W. Topeka Boulevard
Topeka, KS 66611-1287
(785) 274-1401
(785) 274-1426 fax
www.ink.org/public/kdem

**Kentucky Emergency Management**
EOC Building
100 Minuteman Parkway, Bldg. 100
Frankfort, KY 40601-6168
(502) 607-1682
(502) 607-1614 fax
http://kyem.dma.state.ky.us

**Louisiana Office of Emergency Preparedness**
P.O. Box 44217
Baton Rouge, LA 70804
(225) 342-5470
(225) 342-5471 fax
www.loep.state.la.us

**Mainc Emergency Management Agency**
State Office Building, Station 72
Augusta, ME 04333
(207) 626-4503
(207) 626-4499 fax
www.state.me.us/mema/memahome.htm

**CNMI Emergency Management Office**
Office of the Governor
Commonwealth of the Northern Mariana Islands
P.O. Box 10007
Saipan, Mariana Islands 96950
(670) 322-9529
(670) 322-7743 fax
www.cnmiemo.org

**National Disaster Management Office**
Office of the Chief Secretary
P.O. Box 15
Majuro, Republic of the Marshall Islands 96960-0015
(011)(692) 625-5181
(011)(692) 625-6896 fax

**Maryland Emergency Management Agency**
Camp Fretterd Military Reservation
5401 Rue Saint Lo Drive
Reistertown, MD 21136
(410) 517-3600

(877) 636-2872 Toll-Free
(410) 517-3610 fax
www.mema.state.md.us

**Massachusetts Emergency Management Agency**
400 Worcester Road
Framingham, MA 01702-5399
(508) 820-2000
(508) 820-2030 fax
www.state.ma.us/mema

**Michigan Division of Emergency Management**
4000 Collins Road
P.O. Box 30636
Lansing, MI 48909-8136
(517) 333-5042
(517) 333-4987 fax
www.msp.state.mi.us/division/emd/emdweb1.htm

**National Disaster Control Officer**
**Federated States of Micronesia**
P.O. Box PS-53
Kolonia, Pohnpei Micronesia 96941
(011)(691) 320-8815
(011)(691) 320-2785 fax

**Minnesota Division of Emergency Management**
Department of Public Safety
444 Cedar Street, Suite 223
St. Paul, MN 55101-6223
(615) 296-0450
(615) 296-0459 fax
www.dps.state.mn.us/emermgt

**Mississippi Emergency Management Agency**
P.O. Box 4501, Fondren Station
Jackson, MS 39296-4501
(601) 352-9100
(800) 442-6362 Toll Free
(601) 352-8314 fax
www.mema.state.ms.us

**Missouri Emergency Management Agency**
P.O. Box 16
2302 Militia Drive
Jefferson City, MS 65102
(573) 526-9100
(573) 634-7966 fax
www.sema.state.mo.us/semapage.htm

**Montana Division of Disaster & Emergency Services**
1100 North Main
P.O. Box 4789
Helena, MT 59604-4789
(406) 841-3911
(406) 444-3965 fax
www.state.mt.us/dma/des/index.shtml

**Nebraska Emergency Management Agency**
1300 Military Road
Lincoln, NE 68508-1090
(402) 471-7410
(402) 471-7433 fax
www.nebema.org

**Nevada Division of Emergency Management**
2525 South Carson Street
Carson City, NV 89711
(775) 687-4240
(775) 687-6788 fax
http://dem.state.nv.us

**Governor's Office of Emergency Management**
State Office Park South
107 Pleasant Street
Concord, NH 03301
(603) 271-2231
(603) 225-7341 fax

**New Jersey Office of Emergency Management**
P.O. Box 7068
West Trenton, NJ 08628-0068
(609) 538-6050
(609) 538-0345 fax
www.state.nj.us/oem/county

**Emergency Management Bureau**
Department of Public Safety
P.O. Box 1628
13 Bataan Boulevard
Santa Fe, NM 87505
(505) 476-9606
(505) 476-9650
www.dps.nm.org/emc.htm

**New York State Emergency Management Office**
1220 Washington Avenue
Building 22, Suite 101
Albany, NY 12226-2251

(518) 457-2222
(518) 457-9995 fax
www.nysemo.state.ny.us

**North Carolina Division of Emergency Management**
116 West Jones Street
Raleigh, NC 27603
(919) 733-3867
(919) 733-5406 fax
www.dem.dcc.state.nc.us

**North Dakota Division of Emergency Management**
P.O. Box 5511
Bismarck, ND 58506-5511
(701) 328-8100
(701) 328-8181 fax
www.state.nd.us/dem

**Ohio Emergency Management Agency**
2855 W. Dublin Granville Road
Columbus, OH 43235-2206
(614) 889-7150
(614) 889-7183 fax
www.state.oh.us/odps/division/ema

**Office of Civil Emergency Management**
Will Rogers Sequoia Tunnel
2401 N. Lincoln
Oklahoma City, OK 73152
(405) 521-2481
(405) 521-4053 fax
www.odcem.state.ok.us

**Oregon Emergency Management**
Department of State Police
595 Cottage Street NE
Salem, OR 97310
(503) 378-2911 ext. 225
(503) 588-1378
www.osp.state.or.us/oem/oem.htm

**Palau NEMO Coordinator**
Office of the President
P.O. Box 100
Koror, Republic of Palau 96940
(011)(680) 488-2422
(011)(680) 488-3312

**Pennsylvania Emergency Management Agency**
P.O. Box 3321
Harrisburg, PE 17105-3321
(717) 651-2001
(717) 651-2040 fax
www.pema.state.pa.us

**Puerto Rico Emergency Management Agency**
P.O. Box 966597
San Juan, PR 00906-6597
(787) 724-0124
(787) 725-4244 fax

**Rhode Island Emergency Management Agency**
645 New London Ave
Cranston, RI 02920-3003
(401) 946-9996
(401) 944-1891 fax
www.state.ri.us/riema/riemaaa.html

**South Carolina Emergency Management Division**
1100 Fish Hatchery Road
West Columbia, SC 29172
(803) 737-8500
(803) 737-8570 fax
www.state.sc.us/epd

**South Dakota Division of Emergency Management**
500 East Capitol
Pierre, SD 57501-5070
(605) 773-6426
(605) 773-3580 fax
www.state.sd.us/state/executive/military/sddem.htm

**Tennessee Emergency Management Agency**
3041 Sidco Drive
Nashville, TN 37204-1502
(615) 741-4332
(615) 242-9635 fax
www.tnema.org

**Texas Division of Emergency Management**
5805 N. Lamar
Austin, TX 78752
(512) 424-2138
(512) 424-2444 or 7160 fax
www.txdps.state.tx.us/dem

**Utah Division of Comprehensive Emergency Management**
1110 State Office Building
P.O. Box 141710
Salt Lake City, UT 84114-1710
(801) 538-3400
(801) 538-3770 fax
www.cem.ps.state.ut.us

**Vermont Emergency Management Agency**
Department of Public Safety
Waterbury State Complex
103 South Main Street
Waterbury, VT 05671-2101
(802) 244-8721
(802) 244-8655 fax
www.dps.state.vt.us

**Virgin Islands Territorial Emergency Management (VITEMA)**
2-C Contant, A-Q Building
Virgin Islands 00820
(304) 774-2244
(304) 774-1491

**Virginia Department of Emergency Management**
10501 Trade Court
Richmond, VA 23236-3713
(804) 897-6502
(804) 897-6506
www.vdem.state.va.us

**State of Washington Emergency Management Division**
Building 20, M/S: TA-20
Camp Murray, WA 98430-5122
(253) 512-7000
(253) 512-7200 fax
www.wa.gov/wsem

**West Virginia Office of Emergency Services**
Building 1, Room EB-80 1900 Kanawha Boulevard,
East Charleston, WV 25305-0360
(304) 558-5380
(304) 344-4538 fax
www.state.wv.us/wvoes

**Wisconsin Emergency Management**
2400 Wright Street
P.O. Box 7865
Madison, WI 53707-7865

(608) 242-3232
(608) 242-3247 fax
http://badger.state.wi.us/agencies/dma/wem/index.htm

**Wyoming Emergency Management Agency**
5500 Bishop Blvd.
Cheyenne, WY 82009-3320
(307) 777-4920
(307) 635-6017 fax
http://wema.state.wy.us

# Index

# About the Authors

**George D. Haddow** currently serves as an Adjunct Professor at the Institute for Crisis, Disaster, and Risk Management at The George Washington University, Washington, DC. Prior to joining George Washington University, Mr. Haddow worked for eight years in the Office of the Director of the Federal Emergency Management Agency (FEMA) as the White House Liaison and the deputy Chief of Staff. In these positions, Mr. Haddow was involved in the day-to-day management of FEMA, responsible for the Director's communications; policy formulation in the areas of disaster response, public/private partnerships, public information, environmental protection and disaster mitigation including the design and implementation of FEMA's national disaster mitigation initiative entitled Project Impact: Building Disaster Resistant Communities. As the Agency liaison with the White House for Presidential appointments to headquarters and FEMA regional positions, Mr. Haddow worked directly with the FEMA Director and the White House Office of Presidential Personnel in the recruitment and the hiring of all Presidential appointments at FEMA. He also managed FEMA's disaster management and mitigation projects in Argentina, Honduras, El Salvador, Nicaragua, Guatemala, Dominican Republic, Haiti, Ecuador and the Bahamas and coordinated FEMA activities with Korea and South Africa. Served as part of the Clinton Administration's communications team for the Y2K issue.

**Jane A. Bullock** currently serves as an Adjunct Professor at the Institute for Crisis, Disaster, and Risk Management at The George Washington University, Washington, DC. Ms. Bullock has worked in emergency management for over 20 years, most recently as the Chief of Staff to James Lee Witt the Director of the Federal Emergency Management Agency (FEMA). In this position Ms. Bullock served as principal advisor to the Director on all Agency programmatic and administrative activities, provided advice and recommendations to the Director on policies required to carry out the mission of the agency; managed the day-to-day operations of the Agency; directed, monitored, and evaluated Agency strategic and communication processes; and oversaw administration of the Agency's resources, including the disaster relief fund. Represented the Director and the Administration with Congress, State and municipal governments, foreign officials, constituent groups and the media. Served as a principal spokesperson for the Agency's programs both before, during and after disasters. Chief architect of FEMA's Project Impact: Building Disaster Resistant Communities, a nationwide effort by communities and businesses to implement prevention and risk reduction programs. Principal on a project to create National Disaster Response and Mitigation system for Argentina and in six Central American and Caribbean countries. Served as part of the Clinton Administration's Interagency Committee on National Security and Critical Infrastructure Protection. She is the coauthor of a series of books on planning and building communities in hazardous areas.